Measurement of High Temperatures in Furnaces and Processes

David P. DeWitt and
Lyle F. Albright, editors

249 *Volume 82, 1986*

AIChE
Symposium Series

AMERICAN INSTITUTE OF CHEMICAL ENGINEERS

Measurement of High Temperatures in Furnaces and Processes

David P. DeWitt and
Lyle F. Albright, editors

Lyle F. Albright
R. Barber
T.G.R. Beynon
R.P. Bruno
G.J. Burrer
R. Csikos
David P. DeWitt
E.M. Emery
R.L. Grantom
T. Hamazaki
Hoyt C. Hottel
M. Ise
R.P. Madding
James G. Seebold
H. Sugano
K. Welther
I. Zakarias

AIChE Symposium Series

Number 249

1986

Volume 82

Published by

American Institute of Chemical Engineers

345 East 47 Street

New York, New York 10017

Library of Congress Cataloging in Publication Data
Main entry under title:

Measurement of high temperatures in furnaces and processes.

(AIChE symposium series ; no. 249, v. 82)
Includes index.
Papers from a symposium at the National Meeting of the American Institute of Chemical Engineers on Mar. 26–27, 1985, in Houston, Tex. and sponsored by Area 1B (Kinetics, Catalysis, and Reaction Engineering).
1. Chemical processes—Congresses. 2. High temperatures—Measurement—Cogresses. I. DeWitt, David P., 1934–   . II. Albright, Lyle Frederick, 1921–   . III. American Institute of Chemical Engineers. Area 1B. IV. National Meeting of AIChE (1985 : Houston, Tex.)
TP155.7.M42   1986   660.2'8   86–20563
ISBN 0–8169–0363–8

Printed in the United States of America by
Twin Production & Design

FOREWORD

The papers in this publication were presented in a symposium at the National Meeting of the American Institute of Chemical Engineers on March 26 and 27, 1985 in Houston, Texas. The symposium was sponsored by Area 1B (Kinetics, Catalysis, and Reaction Engineering). Information that has become available since the meeting has been added to several papers. All papers were carefully reviewed and edited.

This symposium is the first one in many years devoted to temperature measurements at high temperatures in the petroleum, chemical, and especially petrochemical industries. Some companies have as yet developed little technology in the area. In fact, some companies are still asking the question as to whether there is a need for accurate or even reproducible temperature measurements in ethylene furnaces, dehydrochlorinators in vinyl chloride units, reformers, crude towers, etc. Companies who have investigated matters thoroughly agree that obtaining accurate temperature measurements leads to significantly improved plant operations and to higher profits.

Several papers in this publication discuss the fundamentals of measuring temperatures with radiation thermometers while others consider the application of these thermometers to industrial units. The role and accuracy of thermocouples are discussed in two papers. Radiation thermometers, in conjunction with in-furnace targets, can be used to reliably measure tube wall temperatures under carefully controlled conditions. These targets are provided with preened thermocouples in order to determine accurately the skin temperature of targets. A highlight of the symposium was a panel discussion with subsequent questions from the floor. Chapter 10 is a transcript of the panel discussion.

Limited research to improve temperature measurements is now in progress; and some of it is applicable to the chemical industries. As evidenced by papers of this symposium, numerous suggestions or recommendations have been made and are included in the present publication. Some areas of disagreement were found, and the careful reader will find some in the chapters included here. That point should not serve as a disappointment but rather as a prod to future investigators in their search for the truth and for improved measuring technology.

The editors of the book are especially pleased that the Materials Technology Institute (MTI) located at Columbus, Ohio permitted publication of Chapter 2. This chapter is a condensation of the final report submitted to MTI by the editors, who had made an extensive investigation of the current status of temperature measurements in American chemical industries. Recommendations for the future are also included. We were especially pleased to find that many chemical companies that we interviewed are concerned and wish to develop better measurement techniques since there is considerable economic incentive to do so. With more emphasis in this area, we predict much better techniques will be developed. Hopefully the material in this publication will be an aid in future research and developmental programs.

The editors of this book wish to thank the authors and panelists for excellent presentations and for their cooperation in contributing here. Special thanks go to our secretaries, Mrs. Phyllis Beck (for LFA) and Mrs. Linda Benefield (for DPD) for their extensive efforts in preparing the final manuscripts.

David P. DeWitt
School of Mechanical Engineering
Purdue University
West Lafayette, Indiana

Lyle F. Albright
Department of Chemical Engineering
Purdue University
West Lafayette, Indiana

CONTENTS

BACKGROUND AND PERSPECTIVES ON TEMPERATURE MEASUREMENT IN FURNACES

Hoyt C. Hottel ■ Fuels Research Laboratory, Department of Chemical Engineering
Massachusetts Institute of Technology, Cambridge, Massachusetts

The purpose of this introductory invited lecture is to provide background and perspective on temperature measurements in furnaces. Discussion is presented on interpretation of thermocouple readings and of the different types of apparent radiation temperatures. Methods for measuring luminous flame temperature are reviewed. The importance of understanding the radiative processes within the furnace is especially evident when attempting to measure temperatures.

INTRODUCTION

There are many choices among areas to be considered in introducing the subject of temperature measurement in furnaces. Some material discussed here will, I hope, be new. I shall start with two anecdotes about early experiences in the area, then touch lightly on such matters as the interpretation of furnace thermocouple readings and the case for thermocouple orientation; summarization of blackbody radiation; monochromatic radiation viewed as a power function of temperature, the categorizing of radiation temperature types; two-color pyrometry of luminous flames; the radiating properties of soot in relation to temperature; the use of secondary targets for surface temperature measurement by radiation thermometry; and a review of a few methods of flame temperature measurement. I shall not go into the important subject of how to install thermocouples to measure tube temperatures; I believe that will be adequately covered in coming papers.

About sixty years ago--1925--I spent a year in a steel plant as assistant director of a student practice-school station at the Lackawanna plant of Bethlehem Steel. The students were to run a 48-hour test on a billet-reheating furnace to determine furnace efficiency. My job was partly to supply know-how in areas in which the students' background was inadequate, and I

explained how they would determine the temperature and enthalpy of the billets leaving the furnace. As the billets slid down the exit skids, their temperatures were to be taken with an optical pyrometer. I explained that the billets had an emissivity of about 0.85, and that since the pyrometer was calibrated against a blackbody, a correction would need to be made to the reading. By hindsight, I should have said more. When the test was over, I began studying the data. The readings were much too high, and I asked the test leader how the readings had been taken. With some pride, he explained that the group had discussed my recommended technique, had decided that billets in ambient air were more than 2000°F hotter than their surroundings, but that if they were measured inside the furnace before they hit the skids, they would be surrounded by a temperature much closer to their own, and therefore the reading would be much nearer the correct one. I had not explained to them that optical pyrometers measure an intensity proportional not to the fourth power of temperature but nearer the fourteenth power, and that billets with an emissivity of 0.85 reflected 15 percent of flame-source radiation which was large compared to 85 percent of the radiation from the lower-temperature billet surface. The students had lost the basic data for their 48 hour test! Many of you well know that the point to my anecdote is still very much in the eye of the steel-plant operator today. I shall return to it later.

It was in the middle twenties that engineers began to be very conscious of thermocouple errors in high-temperature systems in which gas and wall temperatures were quite different. For the next decade, papers were written on how to build high-velocity thermocouples, shielded and compensated thermocouples, and--for some applications--silver-covered thermocouples to minimize wall-radiation effects. Furnace men talked about how crude their massive, protected couples were. So, with two students I wrote a paper on temperature measurements in industrial furnaces[1]. When deciding on making a measurement, one should first answer the question "What am I going to do with the information I get?". (If I irritate you by stating the obvious, it is because we are all guilty of not answering all the obvious questions to which we really know the answers!) If in inserting a thermocouple into a furnace chamber one is studying the kinetics of the combustion process, the temperature of interest is of course the true gas temperature, and the thermocouple should be designed accordingly. But that is seldom the interest of the furnace engineer. He is concerned with controlling the heat transfer rate to the stock (or tube) or with the relation of furnace temperature to stock temperature, and the long-life protected thermocouple is far more informative to him than a carefully designed high-velocity thermocouple.

MEASURING HEAT FLUX

For quantitative discussion some terms need defining. Let the blackbody emissive power of a surface of interest--its leaving flux density if black--be E. Let H be the incident flux density and ϵ, α, and ρ the emissivity, absorptivity and reflectance of the surface. The leaving-flux density W will then be ϵE (emitted from the surface) plus ρH (reflected from the surface). The net radiative flux density q_r will then be given by

$$q_r = W - H = \epsilon E - \alpha H = \frac{\epsilon E - \alpha W}{\rho} \qquad (1a,b,c)$$

If the surface is *either* in radiative equilibrium *or* gray, then $\alpha = \epsilon$ and

$$q_r = (\epsilon \text{ or } \alpha)(E - H) = [(\epsilon \text{ or } \alpha)/\rho](E - W) \qquad (2)$$

If the surface is that of a protected thermocouple, with convection much less than radiation, then $q_r = 0$, and from Equation (2) it follows that

$$W \equiv H \equiv E = \sigma T_F^4 \qquad (3)$$

Thus the thermocouple measures a temperature, the "furnace" temperature T_F, from which the incident flux density H is calculable. H is a proper weighting of radiation from gas, refractory and sink, as seen by the thermocouple. If the thermocouple is replaced by a colder body at T_c, the flux to it is given by

$$q = \alpha H - \epsilon E_c = \alpha \sigma \left(T_F^4 - \frac{\epsilon}{\alpha} T_c^4 \right)$$

That is, the measured temperature T_F is the driving temperature to use in computing flux to a heat sink--provided the heat sink has the same view of the furnace interior as the thermocouple.

To eliminate the above constraints associated with shape and position of the thermocouple, let the latter be replaced by the form shown in Figure 1. Two stainless-steel discs are mounted in a porous alundum block, with thermocouples installed to permit determination of either disc temperature or of their temperature difference. If such an instrument, idealized to be free of conduction between plates and convection to gas, were mounted with one disc facing the heat sink and the other the "furnace" as seen by the sink, the hot face would read $E_1 \equiv \sigma T_1^4 = H_F$, the incident flux density from the furnace side; the cold disc would read $E_2 \equiv \sigma T_2^4 = H_s$, the incident flux density from the sink side; and the net flux, furnace-to-sink across the plane of the instrument would be given by

$$q_{net} = H_F - H_s = \sigma(T_1^4 - T_2^4)$$

If allowance is made for the conductive heat leak between the plates and for gas convection (with gas temperature and convection coefficient assumed the same on both sides of the instrument), one has

$$\alpha H_F - \epsilon \sigma T_1^4 + h(T_G - T_1) = \frac{k}{L}(T_1 - T_2)$$

$$= \epsilon \sigma T_2^4 - \alpha H_s - h(T_G - T_2)$$

From this, with $\alpha = \epsilon$,

$$H_F - H_s \equiv q_{net} = \sigma(T_1^4 - T_2^4) \quad (4)$$

$$+ \left[\frac{(2k/L) + h_c}{\epsilon}\right](T_1 - T_2)$$

The bracketed term is an instrument constant which, in the limited calibrating work done, was found to be not quite constant. Experiments in which the hot face was varied between 563 and 913K (553 and 1183°F), the flux density between 15 to 59kW/m² (4630 to 18,600 Btu/ft²hr), and the cold face between 408 to 551K (270 to 530°F) yielded an average instrument constant of 0.0676 kW/m²K (11.9 Btu/ft²hr°F). Use of that value produced an r.m.s. error in predicted flux density of 5.3% or, omitting the lowest flux density, 3.6%. But the time constant of the instrument in the form in which it was tested (3/4 inch of low-conductivity alundum between the two discs) was much too great (ca. 17 minutes). A stack of spaced, thin nickel sheets replacing the alundum would have cured the defect.

BLACKBODY RADIATION

The basic relation in interpretation of furnace radiation measurements is of course the famous Planck's Law giving the monochromatic emissive power of a blackbody E_λ as a function of wavelength and temperature. The origin of quantum mechanics, it is written here in historical reverse to show its relation to its predecessors, the Rayleigh-Jeans and Wien's Laws:

Planck

$$E_\lambda = \frac{2\pi hc^2 n^2 \lambda^{-5}}{\exp\left(\frac{hc}{k\lambda T}\right) - 1} = \frac{c_1 n^2 \lambda^{-5}}{\exp\left(\frac{c_2}{\lambda T}\right) - 1}$$

Rayleigh-Jeans (as $h \rightarrow 0$)

$$E_\lambda = 2\pi ck T n^2 \lambda^{-4}$$

Wien ($\lambda T < 0.3$cm–K)

$$E_\lambda = c_1 n^2 \lambda^{-5} \exp(-c_2/\lambda T)$$

The Rayleigh-Jeans Law, of great historical interest, is correct only in the limit of high values of the wavelength-temperature product, and in consequence has negligible engineering utility. Wien's equation is an excellent approximation to Planck's Law up to $\lambda T = 0.3$ cm–K, and is mathematically more manageable.

Planck's Law may be cast in form to show that E_λ normalized by division by T^5 is a unique function of λT.

$$\frac{E_\lambda}{n^2 T^5} = \frac{2\pi hc^2 (\lambda T)^{-5}}{\exp\left(\frac{c_2}{\lambda T}\right) - 1} = \frac{c_1(\lambda T)^{-5}}{\exp\left(\frac{c_2}{\lambda T}\right) - 1} \quad (5)$$

Figure 2 presents that relation (solid line), together with dashed lines representing the Wien and Rayleigh-Jeans Laws. A scale at the top gives the integral under the Planck curve up to λT, that is, the fraction of blackbody radiation lying below λT. Though the maximum intensity E_λ occurs at 2898 μm–K (Wien's displacement law), the maximum would occur at 5077 μm–K if intensity were defined on a frequency basis E_ν, and two other displacement laws serve the engineer better[2]. One is the value of λT at which the fractional change in E_λ or E_ν per fractional change in λ or ν maximizes, $\lambda T = 3670$ μm–K; the second is the value of λT on either side of which half the blackbody energy lies, 4107 μm–K.

Integration of E_λ over the spectrum gives the total emissive power E. The Stefan-Boltzmann equation was suggested by Stefan's experimental work on platinum radiation and confirmed by Boltzmann's thermodynamic derivation based on using radiation as the working fluid in a piston engine.

$$E \equiv \int_0^\infty E_\lambda d\lambda = \frac{c_1 n^2}{15}\left|\frac{\pi}{c_2}\right|^4 T^4 \equiv \sigma n^2 T^4 \quad (7)$$

The constants of the above equations are summarized here:

$h = 6.6262 \times 10^{-27}$erg sec

$c = 2.997925 \times 10^{12}$cm/sec

$k = 1.38054 \times 10^{-16}$erg/K

$c_1 = 3.7415 \times 10^{-5}$erg cm/sec

$c_2 = 1.4389$ cm K

$\sigma = 5.67 \times 10^{-12}$ W/cm^{2}K$^4 = 56.7 \times 10^{-12}$ kW/m^{2}K^4

$\qquad = 1713 \times 10^{-12}$ Btu/ft^2/hr°R^4

The index of refraction, n, is unity for propagation of radiation through a vacuum.

The intensity of monochromatic radiation, like total radiation, may be thought of as a power function of temperature--over a limited temperature range. The ease of visualizing the consequences of this functional form compared to the Wien or Planck function commends it to the engineer as a tool for easier assessment of sensitivity and errors. The partial derivative of E_λ with respect to T, multiplied by T/E_λ gives

$$\frac{\partial \ln E_\lambda}{\partial \ln T} = \frac{c_2}{\lambda T}\left[\frac{e^{c_2/\lambda T}}{e^{c_2/\lambda T} - 1}\right] \qquad (8)$$

When $\lambda T < 0.3$ cm K, which covers the full range of interest in optical pyrometry as applied to furnaces, the right-hand side reduces to $c_2/\lambda T$. Thus, over a limited but often adequate temperature range, the intensity E_λ is proportional to the $(c_2/\lambda T_{AV})$ power of T, and for optical pyrometry that power is 12 to 15. The above-mentioned problem of reading the temperature of a billet in a furnace is a good example. Consider a billet of emissivity 0.85 at 2200°F (2660°R) inside a furnace at 2600°F (3060°R). Let the Rankin apparent billet temperature reported by the optical pyrometer (red screen, $\lambda = 0.65$ μm) be T_A. Based on the furnace temperature, the power $c_2/\lambda T$ is 14389/0.65(3060/1.8) or 13.02. The 13th-power law gives the approximate T_A' inside the furnace as

$$T_A = (0.85 \times 2660^{13} + 0.15 \times 3060^{13})^{1/13}$$

$$\qquad = 2780°R$$

or 2320°F. Outside the furnace it would be $2660 \times (0.85)^{1/15}$, or 2631°R (2171°F). Done correctly with Wien's law, these numbers would be 2320°F and 2171°F. The power law is mostly for use in thinking, but even numerically it is a good approximation.

APPARENT TEMPERATURES

There are three major classes of apparent temperature measured by radiation instruments:

1) The *total-radiation* apparent temperature T_A is that of a blackbody giving the same total-radiation pyrometer reading as the object sighted on.

$$\sigma T_A^4 = \epsilon_t \sigma T^4$$

or

$$T_A/T = (\epsilon_t)^{1/4}$$

2) The *optical pyrometer* T_A is that of a blackbody having the same red brightness as the object sighted on. Use of Wien's Law gives

$$e^{-c_2/\lambda T_A} = \epsilon_\lambda e^{-c_2/\lambda T}$$

$$\frac{1}{T_A} - \frac{1}{T} = \frac{\lambda}{c_2}\ln\epsilon_\lambda$$

Use of the power-law approximation $c_2/\lambda T_{AV} = n$ gives

$$T_A/T = (\epsilon_\lambda)^{1/n}$$

Any *monochromatic-radiation thermometer* is in this class, with "brightness" replaced by "monochromatic intensity".

3) *Color temperature* is that of a blackbody having the same color as the object sighted on or, to eliminate use of the human eye, the same ratio of intensity at two specified wavelengths. Let two wavelengths be chosen, λ_R and λ_G, and let the measured R/G intensity ratio be B.

$$B = \left[\frac{\lambda_R}{\lambda_G}\right]^{-5} \times \frac{e^{-c_2/\lambda_R T_A}}{e^{-c_2/\lambda_G T_A}} = \frac{\epsilon_R}{\epsilon_G}\left[\frac{\lambda_R}{\lambda_G}\right]^{-5} \times \frac{e^{-c_2/\lambda_R T}}{e^{-c_2/\lambda_G T}}$$

From the first pair of these,

$$T_A = -c_2\left[\frac{1}{\lambda_R} - \frac{1}{\lambda_G}\right]\Big/[\ln B + 5\ln(\lambda_R/\lambda_G)] \qquad (9)$$

From the second pair,

$$T_A/T = 1 + T_A \ln(\epsilon_R/\epsilon_G)/c_2 \left[\frac{1}{\lambda_R} - \frac{1}{\lambda_G} \right] \quad (10)$$

If instead of measuring B or reporting T_A one measures the apparent temperatures $T_{A,R}$ and $T_{A,G}$ at the two wavelengths, T is given by

$$T = \frac{\dfrac{1}{\lambda_R} + \dfrac{1}{\lambda_G}}{\dfrac{1}{c_2} \ln \dfrac{\epsilon_R}{\epsilon_G} + \dfrac{1}{\lambda_R T_{A,R}} - \dfrac{1}{\lambda_G T_{A,G}}} \quad (11)$$

The merit of using color is that for a gray surface of *any* emissivity, the color temperature equals the true temperature.

TWO-COLOR PYROMETRY

Use of the two-color optical pyrometer and the high accuracy and precision possible in optical pyrometry based on a good instrument will be illustrated by reference to work done many years ago[3] on the Hefner candle, an amyl-acetate-burning lamp which in the early days provided the standard candle power of illumination. Six simulated Hefner lamps were set up to permit being moved individually in and out of the line of sight of an optical pyrometer provided with a red and a green screen. Knowledge of the way the absorption coefficient of soot varies with wavelength should permit, in principle, the determination of true temperature T from a pair of apparent red and green temperature T_R and T_G. The objective of the experiment was in part to assess the technique as a way to measure luminous flame temperatures. Before T was computed from any pair of red and green readings, it was first determined by treating all the data on 1 to 6 flame depths at a single wavelength to find the reading for an infinite depth (black flame). The equation for n flames in a row, with k_R the absorption coefficient at λ_R for a single flame, is

$$[1 - \exp(-nk_R)]\exp\left(-\frac{c_2}{\lambda_R T}\right) = \exp\left(-\frac{c_2}{\lambda_R T_{R,n}}\right)$$

or

$$nk_R = -\ln \left[1 - \exp\left[-\frac{c_2}{\lambda_R} \left(\frac{1}{T_{R,n}} - \frac{1}{T} \right) \right] \right] \quad (12)$$

In the pre-computer days one guessed T and plotted log (r.h.s.) versus log n, obtaining a line which, with T guessed correctly, is straight and has a slope of 1 (slope is the better criterion).

The first line of Table 1 shows that the series of red measurements on one to six flame depths yields a temperature of 1696K. The second line shows green measurements, leading to 1694K.

Table 1. Temperature Measurements on Hefner Candle

Flame Depth, n	1	2	3	4	5	6	∞*	Averaged T
T_R, K	1475	1536	1583	1600	1621	1635	1695.7	
								1694.9
T_G, K	1528	1582	1619	1632	1649	1656	1694.0	
T, K*	1696.3	1704.6	1695.5	1693.1	1696.8	1684.9		1695.2

*Calculated value

Comparable agreement was not expected for temperature computed from a pair of red and green readings. It was encouraging to find that n individual T_R:T_G pair of readings predicted a temperature differing more than 10K from the well-established average value of 1695K and that the average of the T's computed from paired readings was within a fraction of a degree of that value. (Air humidity had not been controlled, and at this level of accuracy, it was significant.)

The two-color optical pyrometer used in the above work was later used to determine the temperature of a rather poorly controlled laminar flame of Cambridge city-gas (530 Btu/cu.ft.). The flame was 15 to 40 cm. long, measured at 5 to 20 cm from the burner port. Over a temperature variation of 190K as determined by the two-color method, the temperature averaged 10K above that determined by the Schmidt-Kurlbaum method (see below), with an average variation of 26K (Ref. 4).

LUMINOUS FLAME PROPERTIES

Radiation measurements made in furnace chambers are considerably affected by the flame luminosity due to soot as to justify a summary here of soot radiating properties. These properties depend almost wholly on the product of path length L by mole fraction S of soot in the gas, i.e., on the moles of carbon per mole of gas in which it is suspended, and *not* (or, at least, hardly at all) on gas temperature. Doubling the absolute temperature of a flame in which the soot mole fraction and path length are constant will at constant pressure halve the concentration of soot, but not change the soot *total* emission. This generalization is a great simplifier.

The monochromatic emissivity of a gas due to its soot content *does* depend on temperature T [K], and is given by

$$\epsilon_\lambda = 1 - \exp\left(-\frac{0.526SL}{\lambda T}\right) \qquad (13)$$

There is some disagreement as to the constant, with evidence that it depends somewhat on the age of the soot, the temperature at which it is formed, and its hydrogen content[5,6,7]. The value given is in the middle range of values determined by different investigators. Doubt about the value is far less a matter of concern to the engineer

than the inadequacy of our knowledge of how to predict S. Though there is evidence that the power on λ is not 1 throughout its range. The power (α) is 1.39, not 1, for the wavelength range 0.556 -*0.665 μm of the two-color study of Hefner flames. The energy-important wavelength range is however well beyond 1 μm where there is strong evidence that a power of one is about correct for λ.

Since both T^5-normalized blackbody monochromatic emissive power and soot emissivity are functions of λT, they can be incorporated on the same plot. Figure 3 compares the T^5-normalized emission from a blackbody with the emission from soot having an SL of 0.001713m; such a value is chosen to make the emissivity 0.5 for a 2000K flame viewed at $\lambda = 0.65$ μm. The magnitude of the decrease in emissivity at higher wavelengths is striking. (Additional comparison for orientation: the dotted line is adapted from Rubens' early measurements[8] on a Bunsen burner). Any other luminous flame of the same absorption strength but different temperature would lie on the same curve, but its emissivity at 0.65 μm would be different.

The *total* emissivity due to soot comes from

$$\epsilon_{\text{total}} = \int_0^\infty E_\lambda \epsilon_\lambda d\lambda / \sigma T^4$$

With σ set in from Eq. (7), and ϵ_λ from Eq. (13) and E_λ from Planck's law, the above integrates (with $0.526/c_2 = 36.6$ m^{-1}) to

$$\epsilon_{\text{total}} = 1 - \frac{15}{\pi^4}\left[\psi^{(3)}(1 + 36.6 \text{ SL})\right]$$

where $\psi^{(3)}(1 + x)$ is the pentagamma function of x, available in tables. Fortunately, it may be shown that $(15/\pi^4)\psi^{(3)}(1 + x)$ is equal approximately to $(1 + 0.945x)^{-4}$, with a maximum error of 0.8%. In consequence, with $0.945 \times 36.6 = 34.5$, the total emissivity due to soot is given with adequate accuracy by

$$\epsilon_{\text{total}} = 1 - (1 + 34.5 \text{ SL})^{-4} \qquad [L = m] \quad (14)$$

---

*   The fraction of blackbody radiation at wavelengths below 0.66 μm is less than 1% for a temperature of 2000K. Below 0.55 μm the fraction is less than 0.1%.

The total absorptivity of soot does not equal total emissivity since it depends on the spectral distribution of energy in the radiation source. Let $\alpha_{F,1}$ represent the total absorptivity--due to soot--of a luminous flame at T_F for radiation from a black or gray source at T_1. With 0.526 in Eq. (13), replaced by Kc_2 [$K = 36.6 \text{ m}^{-1}$],

$$\alpha_{F,1} = 1 - \frac{\int_0^\infty \exp\left(\frac{-Kc_2 SL}{\lambda T_F}\right) E_\lambda(T) d\lambda}{\sigma T_1^4} \quad (15)$$

the above becomes

$$\sigma_{F,1} = 1 - \frac{15}{\pi^4} \int_0^\infty \frac{x^{-5} dx}{\exp\left(\frac{KSLT_1/T_F}{x}\right)\left(\exp\left(\frac{1}{x}\right) - 1\right)}$$

$$= 1 - \frac{15}{\pi^4}\left[\psi^{(3)}(1 + KSLT_1/T_F)\right]$$

$$\doteq 1 - (1 + 34.9\ SLT_1/T_F)^{-4} \quad [L = m] \quad (16)$$

To the best of my knowledge, this evaluation of the effect of flame and radiation source temperature on flame absorptivity has never been published. The simple structure of Eq. (16) is a prime example of when to refrain from using a computer to make a numerical integration of Eq. (15). To aid in visualizing the significance of the difference between emissivity and absorptivity, consider the earlier example of a 2000K flame with $SL = 0.001713$ m. With a very luminous flame in the visible range (ca. 0.55 μm), with a total emissivity of 0.21, the absorptivity with 1000K radiation is only 0.11.

USE OF A SECONDARY TARGET IN A COMBUSTION CHAMBER FOR CORRECTION OF APPARENT SINK SURFACE TEMPERATURE

The oil industry's interest in determining the temperature of tubes in processing furnaces started many years ago with thermal cracking units. Extension of the thinking discussed above about the interpretation of protected thermocouples generated the follwoing idea. A MgO block of high diffuse reflectance, placed between the roof tubes of a cracking furnace, would have the same view of the hot gas and refractory surfaces constituting the source of the radiation incident on the tube (heat sink) as the tube itself. Hence, an optical-pyrometer reading of the inserted target would measure the amount of reflected energy which causes difficulty in interpretation of the pyrometer reading when sighted on the tube[9].

Let W_s and W_M represent flux densities leaving the tube surface and the MgO target, H the flux density incident on both surfaces from the flame and refractory, and ϵ and ρ the emissivity and reflectance of the tube surface. Since each W is a combination of emission and reflection, one may write

$$W_s = \epsilon_s E_s + \rho_s H \quad \text{and} \quad W_M = \epsilon_M E_M + \rho_M H$$

Elimination of H between these equation gives

$$\frac{W_s - \epsilon_s E_s}{\rho_s} = \frac{W_M - \epsilon_M E_M}{\rho_M} \quad (17)$$

The right-hand side (r.h.s.) has two possible interpretations. If the target is cold, E_M is negligible and r.h.s. $= W_M/\rho_M$. If the block is in radiative equilibrium, $W_M = E_M$ from Eq. (3), and r.h.s. $= W_M$. Thus, the use of W_M/ρ_M is general if ρ_M is given either its true value or 1, depending on whether operation is with a block newly inserted or in radiative equilibrium.

If the sink surface is assumed gray, Eq. (17) becomes

$$\frac{W_s - (1 - \rho_s)E_s}{\rho_s} = \frac{W_M}{\rho_M} \quad (18)$$

This is the relation used to interpret measurements, and since W_s and W_M are measured and reported as apparent temperatures $T_{s,A}$ and $T_{M,A}$, one must know either the surface reflectance or its temperature T_s to obtain the other. Consider ρ_s known. From Eq. (18),

$$\frac{E_s}{W_s} = \frac{1}{\epsilon_s}\left[1 - \frac{\rho_s}{\rho_M}\frac{W_M}{W_s}\right] \quad (19)$$

Expressed in terms of measured T_A and $T_{M,A}$, Eq. (19) becomes, from Wien's Law

$$\frac{1}{T_s} - \frac{1}{T_{s,A}} = \frac{\lambda}{C_2} \times$$

$$\ln \left[\frac{1 - \dfrac{\rho_s}{\rho_M} \exp\left[- \dfrac{c_2}{\lambda}\left(\dfrac{1}{T_{M,A}} - \dfrac{1}{T_{s,A}} \right) \right]}{\epsilon_s} \right] \quad (20)$$

An example: $T_{s,A} = 1490K$, $T_{M,A} = 1535K$, $\epsilon_s = 0.85$, $\rho_M = 0.95$, and $\lambda = 0.65\ \mu m$. Eq. (20) yields $T_s = 1478K$; the power law based on $c_2/\lambda(1490 + 1535)/2 = 14.6$ would give a T_s only 0.2K lower. The effect of errors is of interest. The human eye can identify differences in brightness of one-half to one percent. A two percent error in measuring intensity W_s would raise $T_{s,A}$ by 2K, and that would increase the calculated T_s by 4K. Use of $E_s = 0.75$ instead of 0.85 would lower the calculated T_s by 11K. The effect of this error in ϵ_s would be greater if the billet and furnace temperatures differed more.

The above analysis of the problem of using an auxiliary target has omitted several factors. These include allowance for signal attenuation by passage through combustion products, the merits of operation in the CO_2:H_2O spectrum at a wavelength where a window exists--somewhat smudged by the presence of any soot; estimation of the relation among flame, refractory, and sink temperatures; and variation of refractory emissivity with wavelength. A rigorous treatment may well be so cumbersome and of such limited validity as to be of less value than the results of an experimental program guided only somewhat by theory.

SCHMIDT-KURLBAUM METHOD AND MODIFICATION

The mean temperature of a gas or flame mass along the line of sight of a radiation-measuring instrument may be determined with confidence if a controlled-temperature background can be placed behind the flame. First used by Kurlbaum and further developed by H. Schmidt, the method depends on adjusting the background temperature until the difference between the instrument readings sighted directly on the background and through the flame to the background disappears. The background if black is then at flame temperature.

Let E_B and E_F be the emissive powers of the black background and of a blackbody at flame temperature, ϵ_F and α_F the emissivity and absorptivity of the flame, and W_F, W_B, and W_{FB} the flux densities leaving the flame, the background, and the background seen through the flame. One may write

$$W_F = \epsilon_F E_F \quad (21)$$

and

$$W_F = E_F \epsilon_F + E_B(1 - \alpha_F)$$

The condition for a match of the two readings on the background is that $W_F = E_B$. Equation (2) then gives

$$E_F \epsilon_F = E_B \alpha_F \quad (22)$$

The system at match is in radiative equilibrium when $\epsilon_F = \alpha_F$; so $E_B = E_F$. Whether a total-radiation pyrometer or a monochromatic radiation thermometer is used is immaterial in principle; a red-screen optical pyrometer sighted on an almost non-luminous flame would however receive little flame-generated radiation and the accuracy of the method would be low. Furthermore if the flame temperature varied along the line of sight, α_F/ϵ_F would vary about 1 in a way dependent on the absorption coefficient and therefore on the wavelength chosen.

The literature is full of examples of effective use of the Schmidt-Kurlbaum method. Need for a peephole on the backside of the flame where a controlled blackbody radiator can be mounted is, however, a serious drawback to use of the method in an industrial furnace. A variation known as the modified Schmidt method is to substitute the furnace backwall as the background. This has had extensive use, but realization of the error associated with the resultant non-equilibrium radiation is limited. Let the nomenclature be the same as above, except that E_B is replaced by W_B, the flux density leaving the non-black backwall. The reflectance of the latter must be known; let it be $\rho_B(\equiv 1 - \alpha_B)$, and assume the backwall gray, i.e., $\epsilon_B = \alpha_B = 1 - \rho_B$. Equation (21) still applies, but Eq. (22) becomes

$$W_{FB} = W_F[1 + \rho_B(1 - \alpha_{F,F})] + W_B(1 - \alpha_{F,B}) \quad (23)$$

There are now two flame absorptances; $\alpha_{F,F}$ is that for flame-source radiation reflected from the backwall (unimportantly small if ρ_B is small) and $\alpha_{F,B}$ is that for radiation emitted by the backwall. The backwall reading W_B is taken with the flame momentarily turned off.

From here on, assumptions may introduce errors. The first is that the two α_F's are identical; one may then solve Eq. (23) for α_F:

$$\alpha_F = 1 - \frac{W_{FB}/W_F - 1}{\rho_B + W_B/W_F}$$

The second assumption is that $\alpha_F = \epsilon_F$; one may then combine the above with Eq. (21) to find E_F:

$$E_F = \frac{W_F(\rho_B + W_B/W_F)}{1 + \rho_B + (W_B/W_F) - (W_{FB}/W_F)} \quad (24)$$

If the radiation measurements are monochromatic, the assumptions made are valid, but if total radiation is used the errors can be large.

To illustrate the details of use of the modified Schmidt method and the magnitude of errors associated with its misuse, an example will be given based on use both of a total-radiation pyrometer and a monochromatic instrument. A red-screen optical pyrometer is chosen as representative of the latter. Though the instrument readings given to illustrate the method are fictitious, they have been carefully constructed from knowledge of CO_2, H_2O and soot radiation. They consider how absorptance is affected by the character of the radiation source, and they are thought to be quite realistic. First, consider use of an optical pyrometer on a somewhat luminous flame. Consider the following apparent temperatures:

flame, $T_{F,A}$ (measuring W_F) = 1824K
flame-plus-backwall, T_{FB} (measuring W_{FB})=1861K
backwall, $T_{B,A}$ (measuring W_B) = 1655K

and let the backwall be gray, with reflectance $\rho_B = 0.3$. From Wien's Law any monochromatic flux ratio W_1/W_2 is expressable as $\exp[-(c_2/\lambda)(1/T_1 - 1/T_2)]$, from which Eq. (24) gives

$$\frac{1}{T_F} = \frac{1}{T_{F,A}} - \frac{\lambda}{c_2} \times \quad (25)$$

$$\ln\left\{ \left[\rho_B + \exp[-(c_2/\lambda)(1/T_{B,A} - 1/T_{F,A})]\right] \middle/ \right.$$

$$\left[1 + \rho_B + \exp[-(c_2/\lambda)(1/T_{B,A} - 1/T_{F,A})]\right]$$

$$\left. - \exp[-(c_2/\lambda)(1/T_{FB} - T_{F,A})]\right]\right\}$$

If the power-law approximation is used instead, with T_{av} in $c_2/\lambda T_{av}$ taken as the arithmetic mean of the three measured T's, the power is 12.4 and from Eq. (24)

$$T_F = T_{F,A} \quad (25a)$$

$$\left(\frac{\rho_B + (T_{B,A}/T_{F,A})^{12.4}}{1 + \rho_B + (T_{B,A}/T_{F,A})^{12.4} - (T_{FB}/T_{F,A})^{12.4}}\right)^{1/12.4}$$

The first formulation gives $T_F = 1922K$, the power-law approximation 1920K. An intermediate calculation (the bracket in Eq. (25a)) gives $\epsilon_F = 0.537$ at $\lambda = 0.65\ \mu$m. That may be used to obtain the soot absorption strength SL. From Eq. (13), SL = 0.00183 m, and from Eq. (14), the soot *total* emissivity is 0.23. If the flame temperature does not vary too much along its line of sight, the temperature determined, 1922K, should be valid.

Now consider use of a total-radiation pyrometer on the same flame. The new data are:

$T_{F,A} = 1531K$ $T_{FB} = 1769K$ $T_{B,A} = 1555K$

Use of the Stefan-Boltzmann law to interpret Eq. (24) gives

$$T_F = T_{F,A} \times$$

$$\left[\frac{\rho_B + (T_{B,A}/T_{F,A})^4}{1 + \rho_B + (T_{B,A}/T_{F,A})^4 - (T_{FB}/T_{F,A})^4}\right]^{1/4} \quad (26)$$

$$T_F = 1895K$$

The inequality of flame absorptance for flame radiation, flame absorptance for backwall radiation, and flame emissivity has introduced an error of 27K in the determination of flame temperature by the modified Schmidt method when the sensing instrument is a total-radiation pyrometer. In spite of the error, there is some merit in using that instrument for the limited objective of assessing *changes* in operating conditions. Such an instrument does a better job of averaging over a line of sight along which the temperature variation is large. One way of visualizing that effect: it can be shown that when the emissive power along the line of sight of a radiation-measuring instrument is linear in distance and the line of sight extends without limit, the averaged emissive power reported by the instrument is the local value at one mean-free-path of the photons, when measured from the near end of the line of sight; the mean free path is the reciprocal of the absorption coefficient. When total radiation is measured, there is a wide distribution of absorption coefficients and the signal received is representative of a less localized source.

MIRROR BACKGROUND

A variation on the modified Schmidt method is the use of a mirror as background. With W_F representing the flux density from the flame alone and W_{FM} that from the flame with mirror behind it, and with ρ_M the mirror reflectance, one has, in addition to Eq. (21),

$$W_{FM} = W_F[1 + \rho_M(1 - \alpha_F)]$$

The assumption that $\alpha_F = \epsilon_F$ then gives

$$E_F = W_F \frac{\rho_M}{1 + \rho_M - W_{FM}/W_F} \qquad (27)$$

Wien's Law applied to Eq. (27) gives

$$T_F = 1 \Big/ \left\{ \frac{1}{T_{F,A}} + \frac{\lambda}{c_2} \ln\left\{(1 + \rho_M - \exp[-(c_2/\lambda)(1/T_{FM} - 1/T_{F,A})])/\rho_M\right\} \right\} \qquad (28)$$

The flame used in the last example would give optical pyrometer readings, for a mirror with a reflectance of 0.92, of

$$T_{F,A} = 1824K \qquad T_{FM} = 1879K$$

With this, Eq. (28) gives $T_F = 1922K$, as expected. The simpler power law is

$$T_F = T_{F,A}\left[\frac{\rho_M}{1 + \rho_M - (T_{FM}/T_{F,A})^{12}}\right]^{1/12}$$

which also gives $T_F = 1922K$.

If a total-radiation pyrometer and mirror background method were used on the same flame, the readings would be

$$T_{F,A} = 1531K \qquad T_{FM} = 1687K$$

Use of the Stefan-Boltzmann law to evaluate Eq. (27) gives

$$T_F = T_{F,A}\left[\frac{\rho_M}{1 + \rho_M - (T_{FM}/T_{F,A})^4}\right]^{1/4}$$

Insertion of observed temperature yields $T_F = 1835K$, 87K in error. For total-radiation pyrometer use of the modified Schmidt method appears to produce less error in gas temperature measurement than the background-mirror method.

The equivalent of the Schmidt-Kurlbaum method and optical pyrometer applied to luminous flames is the sodium-D line reversal method applied to sodium-marked non-luminous flames.[10] With sodium added as a salt spray or as vapor from a molten sodium chloride puddle, a spectroscope is sighted through the flame onto a tungsten strip background lamp. The sodium-D line in the spectrum will appear dark or bright, depending on whether the yellow-brightness temperature of the lamp is below or above true flame temperature. If the lamp has been calibrated with an optical pyrometer, the red and yellow emissivities of tungsten must be used to convert the calibration curve to one based on yellow rather than red brightness. In application of the method to industrial furnace flames, care must be taken to minimize sodium concentration in the cold gas near the walls; the sodium there acts as a net absorber.

CONCLUSIONS

It should be clear that radiation is an important tool of the engineer in making temperature measurements in furnaces. The chance of obtaining valuable data is greatly improved by an understanding of the radiative processes that occur in the furnaces and on the radiative properties of the heat sources and sinks: the emissivity and absorptivity of the CO_2, H_2O and soot in the gas; the windows in the gas spectrum where an instrument sees mostly backwalls; and the emissivity and absorptivity of the sink surfaces.

REFERENCES

1. Hottel, H.C., Meyer, E.W. and Stewart, I., "Temperature in Industrial Furnaces," *Ind. & Eng. Chem.*" *28*, 710 (1936).

2. Hottel, H.C. and Sarofim, A.F., *Radiative Transfer,* p. 15, McGraw-Hill, New York (1967).

3. Hottel, H.C. and Broughton, F.P., "Determination of True Temperature and Total Radiation from Luminous Gas Flames," *Ind. & Eng. Chem., Anal. Ed., 4,* 166-174 (1932).

4. Cooper, M.A., Unpublished report to H.C. Hottel (1932).

5. Dalzell, W.H. and Sarofim, A.F., *J. Heat Transfer, 91,* 100 (1969).

6. Sarofim, A.F. and Hottel, H.C., 6th Int'l Heat Transfer Conference, Toronto, Ont., August 7, 1978.

7. Hottel, H.C. and Sarofim, A.F., "Radiant Heat Transmission" to be published in *Marks' Mech. Engrs. Hbk.*, 9th Ed.

8. Rubens, H., "Concerning the Emission Spectrum of the Auer Burner," *Annalen der Physik*, IV, 18 (1906).

9. Hottel, H.C., Notes on Furnace Design, MIT Course 10.74 (1962).

10. Griffiths and Awberry, *Proc. Roy. Soc.*, A123, 401 (1929).

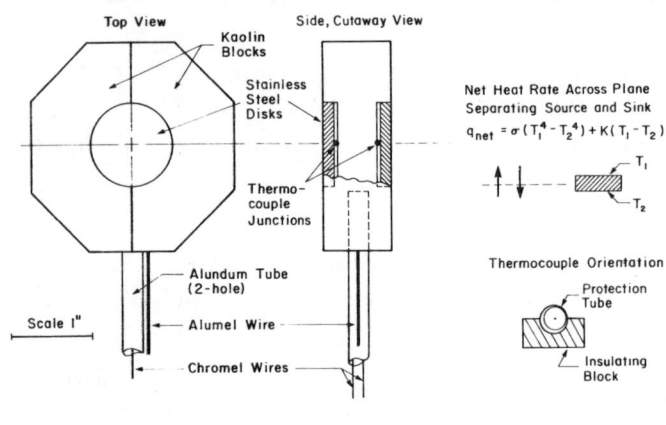

Figure 1. Differential thermocouple as a net flux-density measurer.

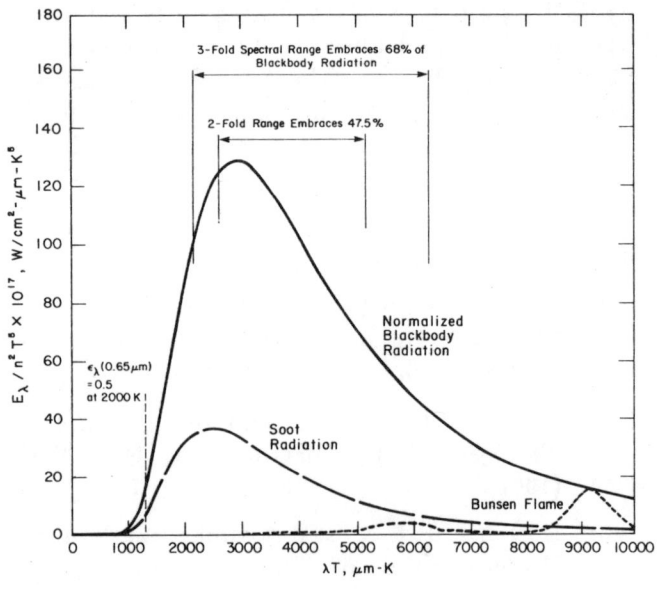

Figure 3. Comparison of Blackbody and Soot Radiation for Absorption Strength SL = 0.001713m having ϵ_{total} = 0.208. Inclusion of Rubens Measurements[8] on Bunsen flame.

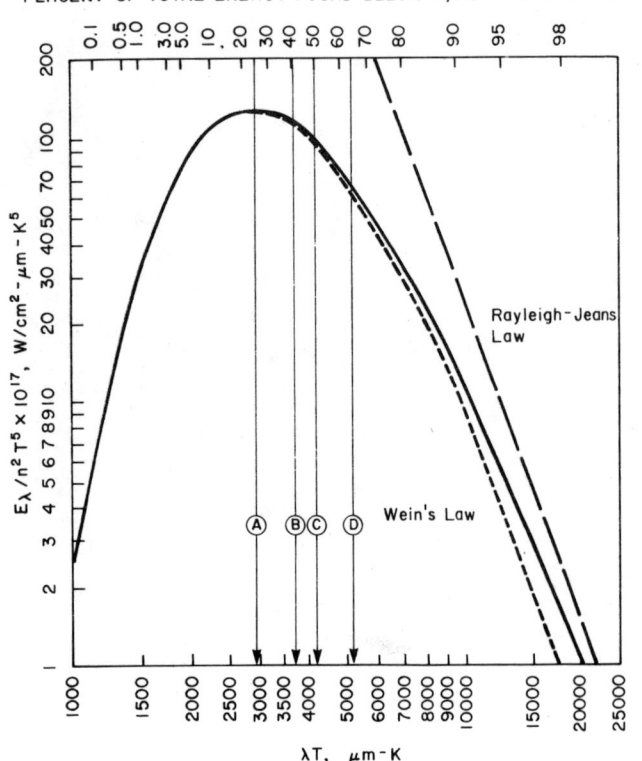

Figure 2. Spectral Properties of Blackbody Radiation. Comparison of Planck, Rayleigh-Jeans, and Wien Laws. Displacement laws where (A) E_λ is a maximum, (B) $\partial(\ln E_\lambda)/\partial(\ln\lambda)$ or $\partial(\ln E_\nu)/\partial(\ln\nu)$ is a maximum, (C) one half energy is on either side of λT, and (D) E_ν is a maximum. Scale of percent of total energy below λT.

TUBE-WALL TEMPERATURE MEASUREMENTS IN FIRED-PROCESS HEATERS: SURVEY OF EXISTING TECHNIQUES AND RECOMMENDATIONS

David P. DeWitt and Lyle F. Albright ■ Schools of Engineering, Purdue University, West Lafayette, IN 47907

Methods of measuring the temperatures of tube walls in fired-process heaters using radiation thermometers and thermocouples are reviewed. Current plant practices are assessed, and recommendations are proposed for further development of the measuring technology.

INTRODUCTION

Measuring tube-wall temperature carefully is of major importance in most, if not all, direct-fired heaters; yet tube life is often significantly reduced by relatively small levels of overheating. For example, Jaske, Simonen, and Roach (1) have discussed tube life in a typical reformer operated at about 500 psia and 871°C (1600°F). A 28°C (50°F) rise in the temperature of the tube results in a 40% reduction in tube life for modified HP alloys and about 100% reduction of HK-40 alloys. Even greater reductions occur with higher temperature rises. Yields of desired products are also often adversely affected by incorrect temperatures. Heiman (2) reports that for an ethylene furnace using a naphtha feedstock and with a rated capacity of about 500,000 metric tons/year, an annual loss of about $1.5 million occurs with a 10°C deviation from the desired outlet temperature. When the temperature is too low, the conversion is too low with less than desired production of ethylene. When the temperature is too high, overcracking occurs with excess undesired by-products including coke are produced.

In 1972, a group at Battelle's Columbus Laboratory (3) made a survey of temperature-measuring techniques in furnaces, of available thermometers, and various temperature-measuring equipment, and of training programs for operators and technical personnel. The group concluded that:

- Temperatures could be measured accurately to within about 14°C (25°F) with available equipment and with well-trained operators.

- Temperatures could be measured to within 8°C (15°F) if tube emissivities, reflected flux, and combustion product effects were better understood.

- Inadequate training and poor understanding of techniques were present in many industrial plants resulting in quite inaccurate temperature measurements.

Based upon the study, the following recommendations were offered: in-furnace probes equipped with calibrated thermocouples should be used to correlate radiation thermometers; surface thermocouples should be installed at selected positions on tubes to supplement and correlate radiation thermometers observations; infrared photography should be used to observe spatial variations along the furnace tube lengths; models of furnaces based on flows, temperatures, conversions, etc. should be developed to infer tube surface temperatures; and spectral radiance variations (rather than just single wavelength) should be observed; and such information should be correlated with other furnace conditions.

The present investigation sponsored by Materials Technology Institute of Columbus, Ohio had as its objectives the following: survey as of 1984-85 of current measurement practices as used in industry; review of progress made since the Battelle survey of 1972; analysis of available instruments for high-temperature measurements; and recommendations for the future. The results of this investigation were reported in detail in the report by D.P. DeWitt and L.F. Albright, "Tube-Wall Temperature Measurements in Fired-Process Heaters", MTI Project #31 Report, 152 pp, 1985. The main features of this report are given here.

INSTRUMENTATION DEVELOPED IN THE LAST 15 YEARS

Nutter (4,5) summarized the status of radiation thermometers available as of 1970 or the approximate time of the Battelle report. Since that time, the following major advances relevant to radiation thermometry have occurred:

- Highly stable silicon photodiode-based radiation thermometers have been developed for operation in the visible and near infrared range (0.6 to 0.9 μm).

- Improved opto-electronic detectors and amplifiers are now available for reliable radiation thermometers that can be satisfactorily operated at any spectral and band-width condition desired from 0.6 to 14 μm.

- Direct-readout of temperature can be obtained as a result of digital circuitry.

- Electronic means are available to adjust for the target emissivity assuming it can be determined.

- Microprocessors plus improved algorithms permit considerations of more process variables in calculating the temperature. Variables that can be incorporated include dual wavelength measurements and furnace wall temperature.

Advances have also been made in contact-type sensors for temperature measurements. Chrome-alumel thermocouples (Type K thermocouples) that have been widely used for many years (6,7,8) often change calibration with time; temperature readings drift within several months by as much as 20°C (36°F) when heated to 800°C or more (2). Improved thermocouples are now finding use at such high temperatures: nicrosil (Ni-14.2% Cr-1.4% Si) versus nisil (Ni-4.4% Si-0.1% Mg). These newer thermocouples have higher thermoelectric stability in air and better oxidation resistance in the range of 1830 to 2190°F (1000 to 1200°C). They generally have much less drift of thermoelectric output (and of predicted temperatures) as compared to Type K (9,10).

MEASUREMENT PROBLEMS WITH RADIATION THERMOMETERS

The key advantages of radiation thermometers relative to thermocouples attached to, or embedded in, the solid walls of a tube or coil include:

- Radiation thermometers do not change the conditions at or near the solid surface whereas thermocouples often shield the surface and/or alter heat transfer to or from the surface.

- Radiation thermometers can measure temperatures at a very large number of locations in a furnace whereas a thermocouple can measure only one.

While radiation thermocouples are potentially highly effective and relatively accurate, they require careful consideration of several operating variables plus a trained operator. Figure 1 illustrates the problems associated with inferring the tube wall temperature, T, using a radiation thermometer. The spectral radiation reaching the thermometer includes the tube wall emission, reflected irradiance (from the refractory walls and other tubes and from combustion products/flames), and absorption/emission of the furnace atmosphere.

A radiation thermometer is calibrated to indicate correctly the temperature of a blackbody. Planck's Law (11,12) is employed to relate the temperature (T) of a *blackbody* (without any line-of-sight atmospheric effects) from spectral radiance, $L_{\lambda,b}$.

$$L_{\lambda,b}(\lambda_{eff},T) = c_1 \lambda^{-5} [\exp\left(\frac{c_2}{\lambda_{eff}T}\right)-1]^{-1} \quad (1)$$

where λ_{eff} is the effective wavelength of the radiation thermometer which is assumed to have a narrow (spectral) bandpass. For this unique situation, inference of temperature from the observed spectral radiance is directly made.

When the radiation thermometer views a tube within a furnace as depicted in Figure 1, the spectral radiance L_λ consists of several radiation components, and inferring tube wall temperature from the indicated temperature is difficult. The simplified situation depicted in Figure 2 is now considered wherein the temperature of the surroundings are as follows: refractory wall, T_w; the flame, T_f; and the tube bank, T_t. Irradiance from the surroundings onto the target is reflected into the field-of-view of the thermometer. The spectral radiance sensed by the detector $L_\lambda(\lambda,T_\lambda)$ is the sum of the radiances due to emission (self-exitance) and reflection,

$$L_\lambda(\lambda,T_\lambda) = \epsilon_\lambda\, L_{\lambda,b}(\lambda,T) + L_{\lambda,ref} \qquad (2)$$

where the reflected radiance is of the form

$$L_{\lambda,ref} = (1-\epsilon)[F_{t-w}L_{\lambda,b}(\lambda,T_w) + F_{t-f}L_{\lambda,b}(\lambda,T_f)$$
$$+ (1-F_{t-w}-F_{t-f})(L_{\lambda,b}(\lambda,T_t)] \qquad (3)$$

The F_{i-j} terms represent the view or configuration factor between the i-j surfaces. Implicit to this analysis is the assumption that the surroundings are isothermal and diffuse behaving as blackbodies. These are generally reasonable assumptions for a first-order analysis.

The utility of Equation (2), hereafter referred to as the *measurement equation,* is illustrated when the furnace temperatures are $T_w = 1065°C$ (1950°F), $T_f = 1650°C$ (3000°F), and $T_t = 955°C$ (1750°F). Realistic values of the view factors can be selected; $F_{t-w} = 0.10$ and $F_{t-f} = 0.05$. That is, 10 and 5% of the reflected flux originates from the wall and flame, respectively, while the balance is from the tube bank. For a typical tube wall emissivity of 0.85, and a radiation thermometer with an effective wavelength of 0.9 μm, the apparent or indicated temperature of 1016°C (1861°F) is high by 61°C (111°F) as compared to the actual temperature, $T_t = T = 955°C$ (1750°F). While the results are based on several assumptions, the measurement equation nevertheless provides

insight into the relative importance of various conditions within the furnace. For example, if the tube wall emissivity, ϵ, was 0.9 (instead of 0.85), the indicated temperature would be 999°C (1830°F), a decrease of 17°C (31°F). A parameter sensitivity study can evaluate the influence of the view factors, flame temperature, and radiation thermometer effective wavelength.

The measurement equation, in some form, must be used in interpretation of radiation thermometer observations on real targets. A better understanding of furnace conditions would permit improved assignment of values to ϵ, the view factos, and/or the surrounding temperatures. These variables are, in general, not well understood in most industrial furnaces, and in such cases, radiation thermometers cannot be relied upon to predict the temperature of the tube walls.

LITERATURE REVIEW FOR RADIATION THERMOMETERS

The literature was reviewed to determine current technology for tube wall temperature measurements. Special attention was given to radiation thermometers for refinery and petrochemical applications.

Lenoir, in 1969 (13), recognized the importance of accurately measuring tube wall temperatures. He indicated that while faulty calibrations of pyrometers often cause serious errors (a real problem at that time), the major difficulties stem from reflected irradiance and hot combustion gases. He concluded that reflected irradiance was of less importance at longer wavelengths. In general, radiation thermometers should be selected with effective wavelengths outside the absorption bands of the combustion gases in the furnace. Lenior concluded that "accurate evaluation of tube wall surface temperatures with a pyrometer requires a great deal of thought and analysis." Furthermore, it is still common even today to assume that a "pyrometer is a pyrometer" and to believe the indicated temperature is the true surface temperature.

In 1973, Nicholson (14), commented on the use of the disappearing filament for measuring tube wall temperatures.

Grandfield, in 1978 (15,16), evaluated the effect of reflected irradiance originating from hot furnace walls. A method was presented that involved observations on the adjacent tube bank (or furnace wall) and the tube itself. Grandfield proposed an algorithm similar to the one of Lenoir (13) for estimating the tube temperature. Without specifically explaining the advantage of a long wavelength radiation thermometer, Grandfield recommended a then new IR instrument operating at 3.9 μm.

In 1983, van Oorschot et al. (17) evaluated eight radiation thermometers (RTs) including: a total (1 to 20 μm) RT, six spectral (narrow band or "brightness pyrometers", with spectral bands in the range 0.5 to 4.0 μm)* RTs and a ratio (nominally 1 μm) RT. He also reported advantages of a *ratio radiation thermometer* (also referred to as a two-color pyrometer) which is a dual wavelength, spectral device (effective wavelengths are close, but the bandpasses generally do not overlap). For this latter instrument, the true temperature is determined from the ratio of the radiance at the two wavelengths, and it offers the following advantages: (1) the indicated temperature is the true surface temperature for a freely radiating (no reflected irradiance effect) *gray* target; hence, no knowledge of the spectral emissivity is required to obtain the true temperature; (2) the target need not fill the field-of-view; and (3) the effects due to gases in the line-of-sight are eliminated if their spectral characteristics are identical at the two effective wavelengths (4,18).

Van Oorschot et al. (17) used a "surface pyrometer" or "gold-cup pyrometer" (19,20) to determine the true temperature of the tube wall. This instrument consists of a gold-plated hemisphere which is placed over the target (tube) and an integral, spectral radiation thermometer (RT) (0.5 to 1.1 μm bandpass) views through the hemisphere surface onto the target. Since the effective emissivity of the target covered by the reflecting hemisphere is nearly unity (typically, if $\epsilon = 0.7$, the effective emissivity is 0.95) and, hence, the indicated temperature by the RT is a good estimate of the true temperature. The sens-

ing head of the gold-cup pyrometer is mounted on the end of a long lance (typically 1.5 m) which accommodates water cooling lines to maintain the sensing head at a fixed temperature. The instrument has, however, two shortcomings: (1) it is unwieldy to handle and to aim at the target; one operator is required to position the lance while the second operates the instrument; and (2) as the sensing head is brought into position, radiant flux to the target is blocked and the tube surface cools; hence, a procedure needs to be developed to infer the undisturbed temperature. While limited to targets on tubes close to peep doors, the instrument provides highly accurate tube temperature measurements, to within 10°C (18°F) at 850°C (1560°F).

Van Oorschot et al. (17) made several important conclusions. First, with properly calibrated radiation thermometers, the precision (repeatability) is typically 7°C (13°F) (see, for example, Figure 3 shown later). Second, errors which include systematic errors are 11°C (20°F), are due to calibration, radiation from the tube environment and emissivity compensation. The best performances are obtained by operating in the infrared region as indicated earlier. Although much uncertainty is instrumental in nature, improved results can, generally, be obtained with careful attention to calibration procedures.

The first van Oorschot et al. study (17), in 1983, concluded on a pessimistic note; they reported that it is impossible to compensate for the systematic errors caused by reflected irradiance from the furnace walls. Later in 1983, van Oorschot (21), however, suggested a correction technique using a graphical approach. The mathematical model to evaluate the reflection effect is similar to Equations (2) and (3) (but $F_{t-f} = 0$) and includes six parameters: indicated temperature of the tube target, (average) temperature of the furnace wall, emissivity of the tube, view factor between target and wall (referred to as the geometry correction factor), bandwidth of the radiation thermometer and effective wavelength of the radiation thermometer. The success of the correction method depends upon the accuracy of determining the six parameters. Unfortunately, detailed evidence is omitted so that the quality of the experimental results and conclusions can not be evaluated. This investigation does, however, clarify the

---

* While the specific instruments employed in the study were not identified, apparently a modern, narrow-band silicon photodiode-based thermometer was not included. The study used instead, two thermometers having 0.5 to 1.1 μm bandpasses.

effort and quality of instrumentation required to make state-of-the-art temperature measurements.

A potentially important contribution is the "laser radiometer" (22) developed in 1983. A laser irradiates the target and a detection system with hetrodyning or optical filtering senses the reflected laser radiation for which the target emissivity can be calculated; from measured spectral radiance emitted by the target, the temperature can be determined. Even though the effect of reflected irradiance from the furnace walls, flames, etc. have not been considered, the method is promising for obtaining accurate spectral emissivity values.

A mechanized search* was performed on temperature measurement developments for the period 1970 to the present. Government literature on the subject covered new applications and on-going work which are primarily oriented to defense problems. A new instrument has been developed to measure graphite temperatures in induction furnaces during pressing (23); the automatic sensor replaces manual pyrometers and reduces demands on the operator. A ratio thermometer (two-color pyrometer) has been used in a test facility for measuring tube walls in the radiant section of a coal-fired MHD power plant (24). For temperature measurement of electrode materials for a thermionic converter, a photon-counting radiation thermometer was used (25). Such thermometers have high sensitivity and are often employed in standardizing laboratories (26). A multi-channel optical pyrometer for 1500 to 3000K has been developed to observe simultaneously five targets (27). The influence of polarization on target emission is considered; fiber optical light pipes are used in the construction of the instrument. It was concluded that this technique has the least intrinsic error when compared with multi-spectral techniques (such as the ratio thermometer, etc.). An infrared band detection system has also been developed for determining a burn-out in a nuclear reactor (28) and for the measurement of turbine blade temperatures. The latter problem is confounded by the presence of higher temperature sources and combustion products (29).

---

* The search by Mr. N.C. Katow, Gulf Research and Development was obtained from the INSPEC retrieval system at the Lockheed Corporation based upon NTIS (government report literature) and the Engineering Index data bases.

A broad perspective on current and recent radiation thermometry was published by NITS for the period 1970-1982 (30,31). In addition, several papers describe multispectral methods; such methods were suggested in 1972 by Battelle (3) as promising topics for study. Relative to target emissivity uncertainty, Gardner et al. (32,33,34) have identified limitations and advantages of the multispectral method. Further study is necessary to determine whether this approach would be useful when the target experiences appreciable reflected irradiance. Cashdollar (35) presents a method for measuring the temperature of particulate matter in a flame or explosion; he utilized continuum radiation at three wavelengths removed from the emission bands of the combustion products. Lillquist (36) discusses the problem of measuring surface temperatures in the 700-900K range in combustion gases. Becker (37) has examined the effect of specularity on radiation thermometers that measure hemispherical emissivity of materials. While none of these studies is directly applicable to the measurement of tube wall temperatures, the instrumentation and analyses employed provide useful background.

Progress in radiation thermometry is obvious from the proceedings of the symposia, Temperature, *Its Measurement and Control in Science and Industry* (38,39). The steel industry has provided increased understanding to temperature measurements and to new instrumental developments. Japanese contributions have been especially important. Further improvements can be expected.

Although literature on infrared scanners relevant to tube measurements is limited, activity in this area is increasing. The Thermosense Symposia (40,41), held yearly, provide an overview of developments on scanner technology and numerous applications. Kaplan (41) has surveyed the scanner field and lists available instrumentation plus the names and addresses of suppliers. Additional literature relating to current practice is subsequently identified.

A novel idea for determining tube wall temperature is a fluidic capillary pyrometer (43). It has recently been developed for ultra-high temperature, corrosive/erosive environments such as molten melts or reactor cores. The sensing phenomena is based upon the change in properties with temperature of a fluid passing through a capillary, typically 1 to 2 cm in length. Preliminary evaluations indicate less than 0.5% shift at 1250°C in calibration over 5000 hours in a hearth forge furnace.

PRACTICES WITH RADIATION THERMOMETERS

The authors of the present paper made several plant visits and interviewed numerous plant personnel relative to current practices of tube wall temperature measurements using especially radiation thermometers. First, a summary of instrumentation, including calibration procedures, is presented, followed by discussion on measurement methodology. Finally, the utility of the measurements to plant operations are discussed.

Instrumentation and Calibration

The types of radiation thermometers used in plants vary widely; no single type or manufacturer dominates the field. The spectral ranges, as previously discussed, typically include 0.65 μm (disappearing optical pyrometers), 0.85 μm (silicon photodiode detector-based thermometers), 2.2 μm (lead sulphide detectors) and 3.9 μm (lead selenide detectors). These wavelengths correspond to the windows for combustion products that have minimal effects due to gas emission or absorption.

Extensive use is still made of disappearing-filament pyrometers, even though considerable skill by the operator and time are required to make an observation (as compared to an automatic thermometer). The precision of the observation is furthermore highly dependent upon ambient light level and eye accommodation; hence, observations are frequently made at night to avoid eye strain. Despite these serious shortcomings, its rugged construction and high reliability are especially advantageous for the harsh and hostile plant environment.

Three features of primary importance when choosing an instrument for plant use are as follows. First, the selection of the proper spectral range establishes several other factors: sensitivity and effects of target emissivity and reflected irradiance. Shorter wavelength thermometers in

general provide improved sensitivity. Second, suitable target size at the viewing distance should be chosen such that the tube being observed completely fills the field-of-view. Even though many instruments have reflex sighting capability complete with reticles to identify the target being sensed, not all the radiation reaching the detector originates from the target as it appears to the eye. Third, simplicity and convenience of operation, especially sighting on the target and reading the indicated temperature, are essential. A small, light-weight instrument provides comfort to the technician who must make many observations without tiring.

Unfortunately calibration facilities are generally unavailable at most chemical plants for checking radiation thermometers. Often the thermometer is returned to the manufacturer for servicing and re-certification. The reliability of such a procedure depends upon the frequency of the checks; careful records should be kept to determine if there are any changes in calibration with time. Such recalibrations should often be made every several weeks, especially if the instrument is subjected to harsh use.

A radiation thermometer can be calibrated with high confidence using a high-temperature blackbody furnace as shown in Figure 3. This arrangement permits a direct comparison of a plant radiation thermometer with both a calibrated thermocouple and a reference thermometer (that is independently calibrated by the manufacturer but is never used in the plant). The blackbody furnace (44) needs to be designed with care; otherwise the comparison of the thermometer and thermocouple readings will not be exact, due primarily to temperature non-uniformities in the block (especially in the axial direction) and the low quality of the cavity. These effects are not usually serious since a primary objective is to provide a *reproducible* temperature source. Reproducibility is assured since the working thermometer is calibrated versus the reference thermometer. Experiences suggest that direct comparisons of the thermometers on a monthly or quarterly basis are satisfactory once confidence in the instrumentation is established. Commercial calibration apparatus is available from several manufacturers at less than $10,000, exclusive of the thermometers.

Methodology of the Measurements

Three major aspects in establishing a methodology to obtain reliable observations are as follows:

First, the instrument should thermally soak for about one hour in the environment where it will be used. Furthermore, exposure of the instrument to radiant flux through the peep door should be minimized; use of reflective foil or other type of shielding (also an aid to the operator) is effective. Unfortunately, little attention is often given to the effects caused by moving a thermometer from a relatively dry, cool operating room to the hot, humid environment near the furnace peep holes. Optics may fog, and temperature compensation circuits within the thermometer may not cope with the rapid, large changes. Placing higher demands on the manufacturer's specifications is not a reasonable alternative.

Second, select tube targets that can be viewed at minimum sighting distances to avoid two pitfalls: that of over filling the field-of-view with objects behind the tube and that of minimizing the emission or absorption by combustion products in the line-of-sight. The desirable limits for angle-of-incidence are subject to controversy. Some practitioners claim sighting should be near-normal and not beyond 60° as measured from the surface normal. Others claim high angles-of-incidence do not degrade the observation. Principles for an angle-of-incidence requirements are not easy to establish. Wherever possible, slight off-normal sighting is preferred so that the specularly reflected line-of-sight does not intercept the peep hole or an especially hot or cold region within the furnace. While most tube surfaces are very diffuse, some degree of specularity is undoubtedly present. If a furnace is nearly isothermal (such as an ethylene furnace), far off-normal viewing may be acceptable. For furnaces at lower temperatures and having large temperature differences, high angle-of-incidence viewing should probably be avoided.

Third, the emissivity compensation feature of the radiation thermometer should not be used since missetting the value systematicaly affects all readings. Some practitioners have permanently fixed the setting, to prevent such systematic error. While the spectral emissivities of aged tubes are generally above 0.85 and thought to be gray (independent of wavelength), setting the emissivity compensation may be misleading because of reflected irradiance effects. It is commonly assumed that the predicted temperature with an emissivity setting of 1.0 is more nearly that of the true surface than with the appropriate emissivity value.

The process of selecting targets within the furnace differs with practitioners. Generally, all tubes within a convenient viewing region are surveyed to determine the hottest tubes or non-uniform temperature regions of a tube. In many cases, the tubes most likely to be hottest based on experience or calculations are surveyed. Hot spots on the tubes can indicate unsafe operating temperatures relative to the desired tube life. Since the radiation thermometer observations at most facilities do not provide reliable surface temperatures, interpolation of the results must be made with care and, based on experiences.

Radiation thermometer observations can be effectively utilized when correlated with other furnace parameters. Such correlations indicate when tube wall temperatures are shifting.

With radiation thermometry, an important goal is to establish consistency of the temperature measurements even if the accuracy of measurements are not good. For example, there should be high confidence that if last week's temperatures were 1030°C and this week they are 1040°C, that indeed there has been a 10°C rise. Many plants have accumulated sufficient experience to indicate whether the tubes will withstand the indicated 1040°C for an extended period. Highly repeatable measurements can be obtained through proper attention to instrumentation calibration, good methodology in target selection and correlation with other furnace operating parameters. This practice will provide baseline data to estimate the effects of tube wall emissivity, reflected irradiances, and emission/absorption of gas products.

PRACTICES WITH INFRARED SCANNERS

An infrared scanner is an image-forming radiometer that measures the temperature distribution across a target area referred to as the scene. The most familiar scanner output is provided on a real-time video display, either black/white (gray) or multicolor; imagery can be stored and used later for viewing and/or analysis. Computer-generated digital displays, including corrections for target emissivity variations or other features, are currently available. Scanners generally operate over a wide spectral bandpass, typically 3 to 5 μm or 8 to 14 μm. At higher temperatures, scanners can operate with a very narrow bandpass, typically centered at about 3.9 μm or 10.4 μm. At these wavelengths, the influence of combustion products on the observations is minimized. Commercial scanner instrumentations is limited by comparison to the radiation thermometer market (42).

A scanner can easily obtain temperature differences over a large surface area, whereas with a radiation thermometer, sometimes referred to as a *spot* radiometer, such information is only obtained with great labor. Several firms can make thermal scans, referred to as thermograms, of tube banks, wall sections, or other furnace target areas. Several petrochemical firms have their own instrumentation and perform periodic examinations of selected targets, particularly tube banks, to detect hot spots, changes in the tube heating patterns over extended periods of furnace operations, non-uniformities due to deposits, tube blockages, etc. (45-48). Temperature differences of 1-2°C can be discerned. Claims have been made that readings are accurate to within 25 to 50°C (35 to 90°F) at 1000°C (1832°F), but evidence to substantiate such accuracies is lacking. The same problems of target emissivity, reflected irradiance, and atmospheric absorption/emission that are of concern for radiation thermometers also occur for scanner imagery. Hence, scanner imagery results are no more accurate than those provided by a radiation thermometer.

An added difficulty is that procedures used to calibrate a scanner generally concentrate on providing accurate temperature difference indications rather than on establishing a temperature scale. Unless special attention is given to the calibration procedure, the accuracy of a scanner near 1000°C (1832°F) is generally poorer than that of a radiation thermometer.

To overcome the calibration difficulty, the usual procedure is to use a probe such as illustrated in Figure 4. It is often a section of a tube or flat plate. A thermocouple is installed in the wall of the metal with a sensing junction close to the surface being viewed. The metal material is generally similar in composition and surface finish to the furnace tubes to be observed. The scanner is calibrated as follows: The scanner is set to indicate a selected base-line temperature (typically just below the temperatures expected in the subsequent observations). The probe is inserted through the peep door, and its temperature rise is monitored as a function of time by the scanner. When the scanner indicates the pre-set temperature, the thermocouple output is recorded and this output serves as the base-line temperature for subsequent furnace tube observations. The accuracy of this procedure for establishing the scanner temperature scale has apparently not yet been assessed. A preferred, but more complicated, procedure would be to use a probe with two targets whose steady-state temperatures bracket the expected range to be observed in the furnace. One target is cooled, usually with air, to obtain temperatures approximating those of the tubes, but the other target is uncooled. Preliminary results using this approach have been reported (49,50).

Efforts to compare infrared scanner and radiation thermometer observations of furnace tube targets have not been seriously addressed. There is hence a serious problem when large discrepancies occur between the reported scanner temperature results and the observations from the radiation thermometer used by the plant operators.

PRACTICES WITH THERMOCOUPLE INSTALLATIONS

The junction of the thermocouple when attached to a solid surface affects absorptivity, emissivity, resistances to thermal conduction in the solid surface, and flow patterns of the flowing gases. In addition, some thermal conduction occurs in the thermocouple leads. As a result, the thermocouple readings in many cases do not indicate the temperature of the solid metal surfaces. Improved contact methods promote significant changes in the temperature of the metal at and near the juncture point. A variety of contact methods are used. They include use of metal pad, block, tip, or knife edge. The method of welding the bead to the metal, the geometry provided as the thermocouple approaches the juncture point, and auxiliary equipment (block, pad, retaining clips, or supports) all vary significantly and all affect the thermocouple readings. Two objectives are desired. First, a thermocouple that provides consistent readings with no service problems over extended periods of operation and second, accurate readings. To a considerable extent these objectives are conflicting and both cannot be obtained. Two methods of contact are described next.

For one installation, silver solder or molydisulphide is used to join the sheathed thermocouple to the metal surface; the joint is protected by a 347SS weld pad. The lead wires to the thermocouples are protected by ceramic fish spin insulation (51,52). The metal surface and the thermocouple are well connected, but the weld modifies the temperature of the metal surface.

In another installation as illustrated in Figure 5, a guide tube contacts the side of a reformer tube. This tube protects the thermocouple leads which are cooled by passage of air at a controlled flow rate. This method likely insures excellent contact between the metal and the thermocouple bead (usually Type K). The metal temperature, however, depends upon the air flow rate, and insight is required to determine a flow rate that minimizes thermal disturbances at the junction.

Several features for thermocouple installations are illustrated in Figures 6 and 7 (53). Figure 6 indicates a method of affixing a target plate of known emissivity to the furnace; and temperatures are then measured with Type S thermocouples. By comparing radiation thermometer and thermocouple readings, in-furnace effects can be estimated. Figure 7 shows several methods for attaching thermocouples to insure that proper contact is made without inducing temperature gradients at or near the junction (51,52). Experimental results with these installations have, however, not been reported.

Two accompanying chapters in this book (by Seebold and by Sugano, Hamazaki, and Ize) present extensive experimental information on several thermocouple installations. This information is helpful in "correcting" thermocouple readings.

FIELD EXPERIMENT WITH RADIATION THERMOMETERS

A plant experiment was conducted by the authors using radiation thermometers (RT's) to measure tube wall temperatures in a commercial direct-fired crude oil heater. Data were obtained (1) to demonstrate the effect of radiation thermometer effective wavelength on the indicated tube-wall temperature, (2) to evaluate the effect of reflected irradiance on tube-wall temperature determination, and (3) to obtain experimental data for use in a model accounting for reflected irradiances from flames and refractory furnace walls.

Three types of RT's were used with effective wavelengths of 0.85, 2.2 and 3.9 μm. Without any corrections for in-furnace conditions - tube emissivity, reflected irradiance, and combustion gas absorption/emission - the indicated readings of the RT's decreased from about 1050°C (1925°F) to 860°C (1585°F) with increasing wavelength. This systematic effect is due primarily to reflected irradiance as discussed next.

Experiments with the in-furnace probe consisting of both hot and cold flat surfaces of Cr-Mo material having an emissivity of approximately 0.85. The cold surface (COLD) was maintained at a temperature just below that of the tube wall using air as a coolant. The hot surface (HOT) was uncooled and, hence, at a temperature similar to that of the furnace gases. The probe was inserted 8 feet into the furnace through a peep door. Observations were made with RT's sighted on HOT and COLD. As expected, the indicated readings were systematically lower as the wavelength of the RT's increased. For HOT; the observations for the 3.9 μm RT without any emissivity compensation agreed within 11°C (20°F), of the thermocouple readings. This result was anticipated since the temperatures of HOT and the furnace interior were nearly the same; hence, HOT appeared to the RT as a blackbody. Reflected irradiance had, however, a significant effect on COLD which was at a temperature essentially identical to that of the tube. The predicted temperatures were significantly too high, by values varying from 335°C (600°F) to 135°C (240°F) as the wavelength increased. Clearly the effect of reflected irradiance needs to be evaluated for this furnace. The probe targets should possibly be oriented similar to the tubes, in addition to being in proximity. Unfortunately, physical limitations on the probe construction and peep door size prevented all of the observations desired.

In regards to the third objective, readings were made with two radiation thermometers (0.85 and 3.9 μm effective wavelengths) for two tube locations, adjacent furnace walls, and nearby flames. A model was developed for predicting the tube surface temperature using as input information six indicated RT readings (three observations for each RT: tube, average wall and average flame) and an assumed value for the emissivity of the tube, 0.85. The model provides two equations with two unknowns - the tube surface temperature and the view factor between the tube target and the flame - that can be solved simultaneously. The predicted view factor and tube wall temperatures are reasonable, but without knowledge of the true surface temperature, the accuracy of the model cannot be determined; the model, however, does predict the effects of reflected irradiances from the walls and flames. Certainly even the simplest of models involving in-furnace conditions are complex.

While this experimental investigation is considered highly preliminary, the results indicate the type of information that can be obtained with radiation thermometers and the analyses that can be performed to deduce tube wall temperatures.

RECOMMENDATIONS

The recommendations are divided into three broad categories.

Education

Much improved educational programs are needed in many cases. Many companies lack good educational and training programs for measurement of high temperatures as evidenced by use of old-style instruments, inconsistent data, failure to define clearly measurement problems, and failure of management to provide adequate technical staff and funds. Some companies are, however, notable exceptions with strong programs in place, and they apparently heeded the advice of the Battelle group, made in 1972 (3). The most significant educational shortcoming in most companies is the inadequate technical staff that should have more background in the physics of thermal radiation, selection and use of instruments, and interpretation of the data obtained. Hot or cold spots in a furnace seem common. While the quality of measurements depends on the technician's skill and patience, the plant engineer must decide on the surfaces to be measured and must then interpret the results.

Specifically, companies can, and some have, addressed their measurement problems by establishing task forces of technical and management personnel to assess and define the current status of the problem, and to develop short-term and long-term objectives. Technical personnel should be encouraged to improve their training by literature study, and by attendance at short courses, seminars, and technical meetings. Of course, specialists can also be hired. Third, in-plant training programs for engineers and technicians can be instituted.

Additional Information Required for Radiation Thermometers

Although the basic radiation equations are known, more and better observed data are required for industrial furnaces in order to determine temperature accurately.

Emissivity of Tube Materials: Reliable spectral emissivity data are needed as a function of temperature, materials of construction, and surface composition and roughness. Such information may be available in some companies, but it is not in the open literature. Enough information is, however, probably now available to start a limited data base and to determine the conditions that would be most fruitful to investigate. The following variables need to be carefully considered in the recommended investigation: proper spectral range, environmental conditions similar to those of a furnace, spectral-diffuse nature of the furnace, and surface characteristics.

Spectral Characteristics of Furnace Atmospheres: Considering water vapor-carbon dioxide systems, the emission and absorption are minimal at certain wavelengths, typically 0.65 to 0.9, 1.6, 2.2, 3.9, and 10.6 μm. Two effects need to be considered to develop improved temperature readings. First, there is the line-of-sight absorption phenomenon without the presence of flames. Second, there is the reflected irradiance effect originating from flame radiance; this effect is part of the general problem considered in the previous recommendation.

For furnaces using clean burning gases, particulate matter such as soot is not usually present. The influence of line-of-sight absorption can easily be estimated by sighting through the furnace onto a cold target mounted outside the opposite port to sense the spectral radiance of the hot gases along the line-of-sight. The cold target is then replaced by a blackbody at high and known temperature. The spectral radiance in this case is that due to the blackbody plus the interference of the gas in the line-of-sight path. From knowledge of these two spectral radiances and the path length, the influence of the furnace atmosphere on radiation thermometer observations can be deduced.

For furnaces, such as oil-fired units, where particulate matter is significant, the problem becomes much more complicated. The above experiment can be performed at several wavelengths to determine the preferred spectral condition for minimizing effects. Since scattering is dependent on the particle size, scattering will be wavelength dependent.

Detailed Modeling for Reflected Irradiance: A methodology needs to be developed for diagnosing the importance of irradiance and constructing a model to estimate its importance on radiation thermometer observations. To determine a basis for modeling, extensive data for a well-characterized furnace are needed. In making such a study, appropriate radiation thermometer readings and independent and reliable wall temperature values (as determined by thermocouples, gold-cup pyrometer, and thermodynamic and kinetic evaluations of reacting gas stream) are required. A modeling study could be initiated to estimate the importance of reflected irradiance. The experimental effort should be performed at the furnace using instruments already available; some additional instruments may be needed for special observations. The modeling work would be performed by a group with expertise in radiative transfer analysis.

Improved Thermocouple Technology

Improved understanding of the errors associated with the use of thermocouples should be obtained. Once such errors are better evaluated, then improved methods of using thermocouples will likely become apparent. The information reported in chapters of this book by Seebold and by Sugano, Hamazaki, and Ise will help plan the proposed investigation.

SUMMARY

A combination of radiation thermometers (or infrared scanning devices), thermocouples, calibration devices such as blackbody furnaces, and furnace models can be used to estimate with reasonably good accuracy the skin or surface temperatures of coils (or tubes) in direct-fired furnaces. These models predict the conversion, composition of gaseous products in the coil, and energy release or requirements. This approach requires careful planning and choice of reliable radiation devices and thermocouples. A rather elaborate temperature-measuring program is required at each furnace. Although such a program will require significant cost to develop and then to implement, several companies have found it to be highly cost-effective since it leads to increased yields of products, lower energy requirements, longer run lengths, less maintenance, and longer tube and furnace life. As more data are obtained regarding tube emissivities and furnace gas radiative properties, as better understanding of reflection patterns, and as better equipment is developed, even more accurate temperature measurements can be expected.

LITERATURE CITED

1. Jaske, C.E., F.A. Simonen, and D.B. Roach, "Product Reformer Furnace Tube Life," *Hydrocarbon Processing*, pp. 63-68, January 1983.

2. Heiman, J.C., Economic Considerations of the Design and Operation of Conventional Pyrolysis Furnaces in *Pyrolysis: Theory and Industrial Practice*, (L.F. Albright, B.L. Crynes, and W.H. Corcoran, Editors), Chapt. 14, pp. 365-67 (1983).

3. Linebrick, O.L., P.A. McRury and J.J. Clayton, "Materials for Steam Reformers, Special Task: A Study of Temperature Measurement Problems in Steam Reformer Furnaces", Battelle Special Technical Report, Battelle, Columbus Laboratories, p. 40, 1972. (Proprietary report in 1982 made available through Mr. D.B. Roach, Program Manager).

4. Nutter, G.D., "Radiation Thermometry, Part 1 - Recent Advances and Trends," *Mechanical Engineering,* 94, 12-15, 1972.

5. Nutter, G.D., "Radiation Thermometry, Part 2 - Solving the Emissivity Problem," *Mechanical Engineering,* 94, 16-23, 1972.

6. Kennedy, R.H., "Selecting Temperature Sensors," *Chemical Engineering,* pp. 54-57, Aug. 8, 1983.

7. American National Standards Institute (ANSI), *Temperature Measurement Thermocouples,* Standard MC96.1-1975, Instrument Society of America, Pittsburgh, 1976.

8. American Society for Testing and Materials (ASTM), *Manual on the Use of Thermocouples in Temperature Measurement,* STP 470B, ASTM, Philadelphia, 1981.

9. Burns, G.W., "The Nicrosil versus Nisil Thermocouple: Recent Developments and present Status," in *Temperature, Its Measurement and Control in Science and Industry,* Volume V (J.F. Schooley, Ed.), Part Two, 1121-1127, American Institute of Physics, 1982.

10. Burley, N.A., R.L. Powell, G.W. Burns and M.G. Scroger, *The Nicrosil versus Nisil Thermocouple: Properties and Thermoelectric Reference Data,* National Bureau of Standards, Monograph 161, 1978.

11. Siegel, R. and J.R. Howell, *Thermal Radiation Heat Transfer,* (2nd Edition), McGraw Hill, 1981.

12. Incropera, F.P. and D.P. DeWitt, *Fundamentals of Heat Transfer,* 2nd Edition, Wiley, 1985.

13. Lenoir, J.M., "Furnace Tubes: How Hot," *Hydrocarbon Processing,* pp. 97-101, 1969.

14. Nicholson, R., "Measurement of Tube Metal Temperatures in Radiant Wall Pressures Furnaces Using the Disappearing Filament Pyrometer," *J. Inst. Fuel,* 46 (382) 258-261, 1973.

15. Grandfield, S.D. "Less Error in Radiant Tube Temperature Sensing", Paper MC-78-14, presented at the Refinery and Petrochemical Plant Maintenance Conference, Houston, TX, February 8-10, 1978.

16. Grandfield, S.D., "Method Cuts Error in Radiant Tube Temperature Sensing," *Oil Gas J.,* pp. 68-70, May 1, 1978.

17. Van Oorschot, B.P.J., J.A.M. Hesselman, and W.H. Van den Berg, "Can Brightness Pyrometers Measure Olefins Plant's Furnace Tube Temperatures Accurately," *Oil Gas Journal (Technology),* pp. 194-202, May 2, 1983.

18. Reynolds, P.M., "A Review of Multicolour Pyrometry for Temperatures below 1500° C," *Brit. Journal of Applied Physics,* 15, p. 579, 1964.

19. Drury, H.D. et al. "Pyrometers for Surface Temperature Measurement", *J. Iron Steel Institute,* pp. 245-250, November 1951.

20. Product Data Sheet 57, "Surface Pyrometers and Thermometers", Model QSP Surface Pyrometer, Land Instruments, Inc., Tullytown, PA.

21. Van Oorschot, B.P.J., "Correction of Pyrometer Measurement of Furnace Tube Temperature: A Viable Proposition?" Unpublished manuscript provided by I.A.M. Hesselmann, Koninklijke/Shell Laboratorium, Amsterdam, Sept. 1983.

22. Stein, A., P. Rabinowitz, and A. Kalds, A. (Exxon Research Engineering Co.), "Laser Radiometer," U.S. Pat 4,417,822, (November 29, 1983).

23. Oak Ridge Y-12 Plant, "High-Temperature Carbon-Furnace Thermometer," NTIS Report Order No. DE82001862, January 1983 [987335 NTN83-0029].

24. Mississippi State University, MHD Energy Center, "Testing and Evaluation of MHD Materials and Substructures," Department of Energy Report No. DOE/ET/10785-T1; FE-2246-15, p. 204, June 1981 [81633, DE81024331].

25. Jacobs, M.H. and D. Jacobson, "Work Function Determination of Promising Materials for Thermionic Converters," NASA-CR-164368; ERC-R-7903, p. 24, 1980 [853640, NB1-24526/8].

26. Coates, P.B., B.F. Billing, and T.J. Quinn, "The NPL Photon-Counting Pyrometer," Inst. Phys., Instrumentation and Control, IEE, European Phys. Soc., Temperature Measurement, 238-43, 1975 (NTIS 856696).

27. Pierson, A.H., G. Zeigler, and M. Weiss, "A Multi-Channel Optical Pyrometer for the 1500K to 3000K Range," AFFDL-TR-65-133, p. 81, January 1966 [577360, AD-480487/0].

28. Benn, D.N. and R.A.W. Shock, "Infra-red Burn-out Detectors," Ukaea Research Group, Hornwell, Atomic Energy Research Establishment, AERE-R-7338, p. 10, February 1974 [398048].

29. Buchele, D.R., "Surface Pyrometry in Presence of Radiation from other Sources with Application to Turbine Blade Temperature Measurement," NASA Report NASA-TP-1754; E-396, p. 19, November 1980 [817274, N81-11039/7].

30. National Technical Information Services, "Pyrometers. 1970-August, 1982. Citations from the NTIS Data Base," NTIS Report PB82-872680, p. 145, August 1982, [915999].

31. National Mechanical Information Services, "Pyrometers. 1970-August, 1982. Citations from the Engineering Index Data Base," NTIS Report PB82-87262, p. 245, August 1982, [915998].

32. Gardner, J.L., T.P. Jones, and M.R. Davies, "Six Wavelengths Radiation Pyrometer," *High Temp-High Pressures*, 13, (4), 459-66, 1981.

33. Gardner, J.L., "Computer Modelling of a Multiwavelength Pyrometer for Measuring True Surface Temperature," *High Temp-High Pressures*, 12, (6), 699-705, 1980.

34. Gardner, J.L. and T.P. Jones, "Multi-Wavelength Radiation Thermometry Where Reflectance is Measured to Estimate Emissivity," *J. Phys. E. Sci. Instrum.*, 13, (3), 306-310, 1980.

35. Cashdollar, K.L. "Three Wavelength Pyrometer for Measuring Flame Temperatures", *Applied Optics*, 18, (15), 2595-97, 1979.

36. Lillquist, R.D., "Infrared Pyrometer for Measuring 700-900K Surface Temperatures in a Combustion Gas Background Environment," ASME Paper No. 80-HT-5, American Society of Mechanical Engineers, p. 8, July 1980.

37. Becker, H.B., and T.F. Wall, "Effect of Specular Reflection of Hemispherical Surface Pyrometer on Emissivity Measurement," *J. Phys. E.*, 14, (8), 998-1001, 1980.

38. Plumb, H.H. (Ed.), *Temperature, Its Measurement and Control in Science and Industry*, Volume 4, Part I, Section II, Radiation Thermometry, pp. 377-688, Instr. Soc. Amer., 1972.

39. Schooley, J.F. (Ed.), *Temperature, Its Measurement and Control in Science and Industry*, Volume 5, Part 1 (pp. 1-710) and Part 2 (pp. 711-1395), American Institute of Physics, 1982.

40. Burrer, G.J. Ed., *Thermal Infrared Sensing Diagnostics*, (Thermosense VI), Proc. SPIE 371, 254 pp., 1984.

41. Kantsios, A.G., Ed., *Thermal Infrared Sensing Diagnostics*, (Thermosense VII), Proc. SPIE 520, 229 pp., 1985.

42. Kaplan, H., "An Update of Commercial Infrared Sensing and Imaging Instruments," *Thermal Infrared Sensing Diagnostics*, (Thermosense VI), G.J. Burrer, Editor, Proc. SPIE 371, 239-246, 1984.

43. Negas, T., H.S. Parker, R.M. Phillips, T.M. Drzewiecki and L.P. Dominques, "Fabrication, Testing, and Evaluation of Prototype Fluidic Capillary Pyrometer Systems," *Journal of Dynamic Systems, Measurement and Control,* 103, 309-315, 1981.

44. Bedford, R.E., "Effective Emissivities of Blackbody Cavities -- A Review," *Temperature, Its Measurement and Control in Science and Industry,* Vol. 4, Part 1 (H.H. Plumb, Ed.) pp. 425-434, 1972.

45. Mol, A., "Why Revamp Older Steam Crackers," *Hydrocarbon Processing,* pp. 179-184, July 1978.

46. Hill, A.G. and D.V. Devers, "Infrared Scanning Pays Off for Amoco Refinery," *Oil and Gas Journal,* June 19, 1978.

47. Nielson, C. and J. Powers, "Varied Uses Found in Chemical Plant for Infrared Thermography," *Chemical Processing,* pp. 74-75, April 15, 1979.

48. Ingram, A.G., and J.B. McCandless, "Use Infrared to Find Hot Spots," *Hydrocarbon Processing,* pp. 219-223, May 1982.

49. Ingram, A.G., and J.B. McCandless, "Infrared Thermal Imaging of Refinery Equipment," *Thermal Infrared Sensing Diagnostics,* (Thermosense V), G.E. Courville, Ed., Proc. SPIE 371, 47-54, 1982.

50. Bruno, R.P. and G.J. Burrer, "Appraising Process Furnace Tubes in the Imaging Radiometers," *An International Conference on Thermal Infrared Sensing for Diagnostics and Control,* (Thermosense VI), G.J. Burrer, Ed., Proc. SPIE 446, 130-136, 1984.

51. Finney, P.F. (to Thermo-Couple Products Co., Inc.), "Surface Thermocouple," U.S. Pat. 3,874,239 (April 1, 1975).

52. Finney, P.F. (to Thermo-Couple Products Co., Inc.), "Surface Thermocouple," U.S. Pat. 3,939,554 (February 24, 1976).

53. Vosnick, H.P., Wahl Intruments, Inc., Culver City, CA, Private Communication.

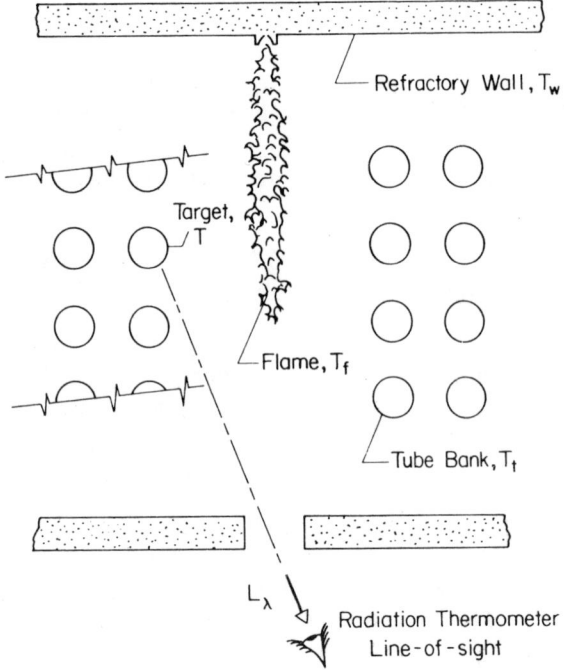

Figure 1. The Measurement Problem: Inferring Tube Wall Temperature T from Spectral Radiance Leaving the Target.

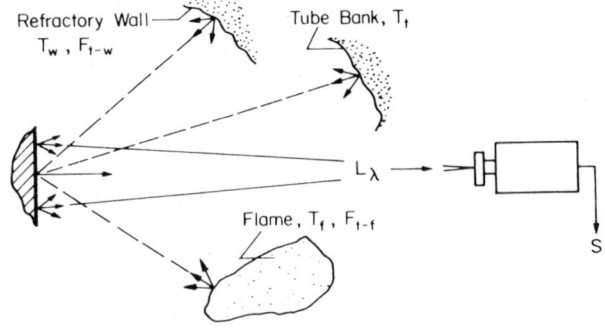

Figure 2. Effect of Reflected Irradiance from Target Surroundings.

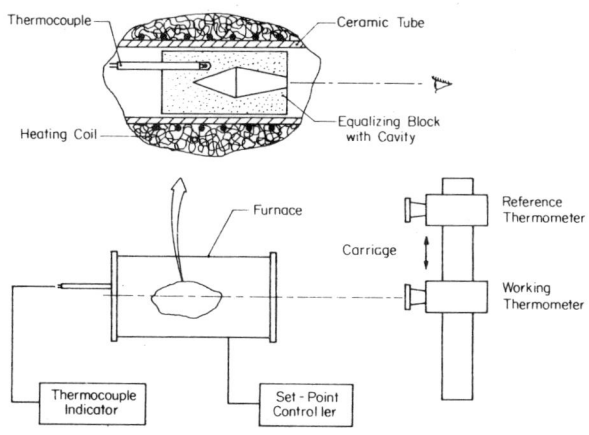

Figure 3. High-Temperature Furnace with Blackbody Cavity for Comparison of Thermocouple and Radiation Thermometer Scales.

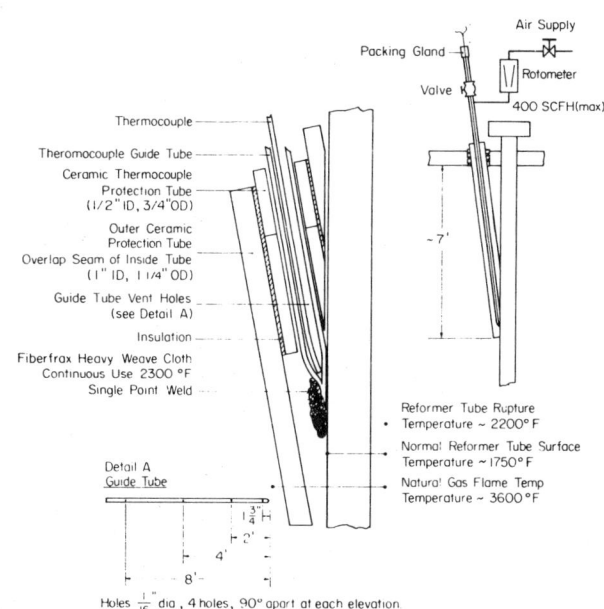

Figure 5. Thermocouple Installation with Air-Purged Guide Tube (Courtesy of Arco Chemical Co.).

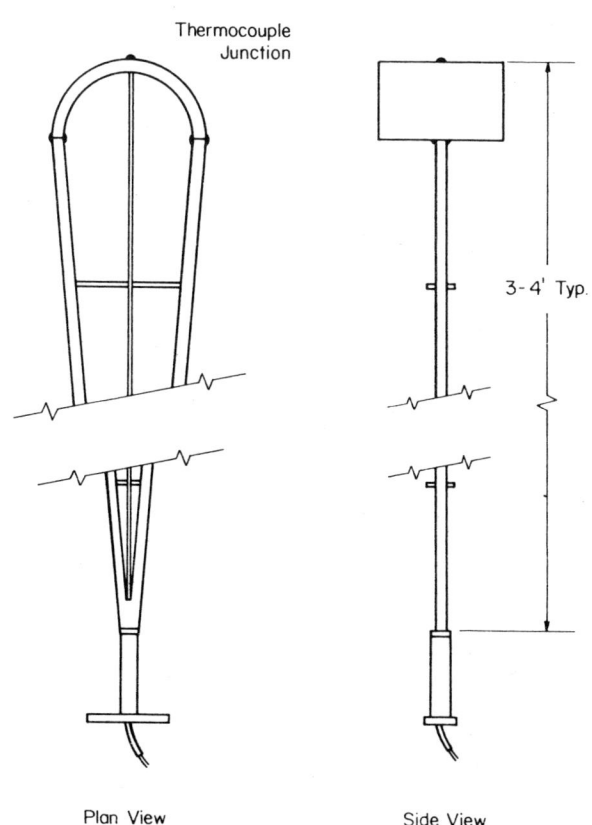

Figure 4. Typical Probe with Thermocoupled Target for Calibration of Infrared Scanners.

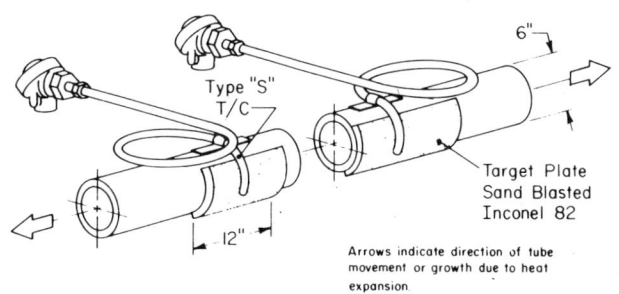

Figure 6. Method of Attaching Thermocouples to Solid Surfaces.

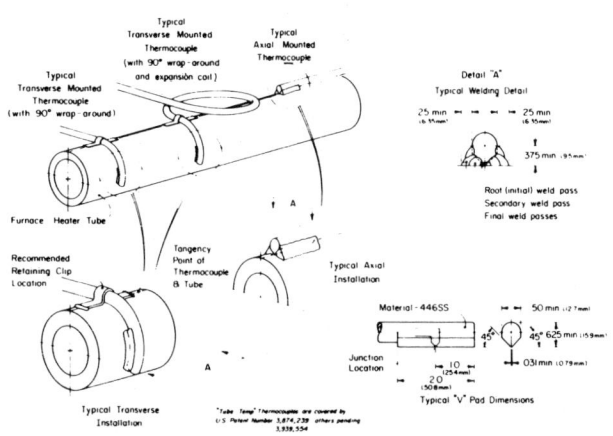

Figure 7. Alternate Methods of Attaching Thermocouples to Solid Surfaces.

A TUBE WALL TEMPERATURE MONITORING PROGRAM

R.L. Grantom ■ E.I. duPont deNemours & Co., Inc., Alvin, Texas 77512

A case example is reported for an ethylene plant on the need for and the development of an accurate temperature-measuring technique. Details of the technique developed are described. Accurate measurement of tube wall temperatures has been found to be important.

INTRODUCTION

A petrochemical plant for producing ethylene and other olefins, diolefins, and aromatics uses fired tubular reactors to steam pyrolyze hydrocarbons. The hydrocarbon feed is first preheated in the convection section; if the feed is a liquid, vaporization occurs. The hot vapor feed then flows through the fired tubular reactors suspended in the radiant section of the heater. These reactors are long, unpacked, serpentine coils hung vertically between the sidewalls of a narrow, refractory lined firebox. The firebox is provided with wall and/or floor mounted burners on both sides of the reactor.

The hydrocarbons being pyrolyzed are frequently heated to 760–885°C (1400–1625°F) in from 0.15 to 0.75 s, depending on the feedstock and the design of the reactor. To achieve the desired heat transfer, heat fluxes as high as 110kW/m^2 (35,000 $\text{Btu/ft}^2/\text{h}$) and tube wall temperatures in the 950–1000°C (1749–1830°F) region are necessary. Unfortunately, during the pyrolysis process, thermal degradation of the feedstock occurs resulting in internal fouling of the reactor by carbon deposits. These deposits increase the resistance to heat transfer through the reactor wall and result in a gradual increase (1–3°C per day) of tube wall temperatures, for the same thermal duty in the reactor. Ultimately, a limiting value of tube wall temperature is approached, necessitating the shutdown of the heater and removal of the internal coke.

Due to high capital costs, spare heater capacity is minimized in an ethylene plant. At full capacity operation, furnace availability may become a problem if furnace on-stream time is not extended to the maximum permitted by tube metallurgy. Since tube wall temperature is the furnace limit most affected by coke deposition, its accurate measurement is essential. Errors of 3–5°C could mean taking a furnace off-line for decoking a day or two before necessary. If this results in the furnaces limiting plant throughput, the economic penalty could be large. Furthermore each time a furnace is decoked, significant costs are incurred for steam, air, operating manpower, and mechanical damage to the furnace.

Excessive tube wall temperatures increase carburization rates of the coil metallurgy resulting in internal metal wastage, and reduction in sound metal thickness of the tube, and early tube failures.

MEASUREMENT METHODS SURVEY

In 1977, Monsanto and Conoco undertook the design and construction of a new, modern high technology ethylene plant at Chocolate Bayou, near Alvin, Texas. Realizing the economic importance of accurate furnace tube wall temperature measurements, a rather extensive program was undertaken to determine the most accurate approach to use. Accuracy was the key word. While precise measurements were desirable they would be of little usefulness if they were in serious error. Several measurement techniques yield consistent, repeatable indications of temperatures but give inaccurate results. In fact, every approach considered, including the approach finally selected required corrections.

Tube skin thermocouples were eliminated from consideration early. Too many would be needed and there was no way of knowing exactly where they should be placed. Due to uneven coke deposition, hot spots often develop along the tubes. Because of the environment, accuracy would be doubtful [1], and horrendous maintenance problems were anticipated. Other contact sensors such as pyrometric cones, coatings, and decals were eliminated for obvious reasons. Only non-contact sensors such as radiation thermometers were left for consideration. Based on previous bad experience with non-contact sensors this was not a consoling thought, but that was the situation faced at that time.

Disappearing optical pyrometers are an old type of non-contact sensors. Their mode of operation is to vary the current flow through the filament of an incandescent lamp, by changing a variable, temperature-calibrated resistor, until a bright spot (the projected lamp filament) visually "disappears" into the hot target background. Optical pyrometers are rugged, portable, and low cost. These are significant advantages. They have, however, a number of disadvantages for measurements of this type. Emissivity compensation can only be estimated with difficulty. Reflected energy from visible light sources (surrounding radiant walls, for example) in the vicinity of the target cannot be distinguished from the target's own self-emitted radiation. The optics do not lend themselves to easily finding a target. Their most serious disadvantage, however, is that a human observer is required. Human error and person-to-person variance impairs accuracy since visual acuity and color blindness may be involved.

Nicholson [2] investigated the use of optical pyrometers for measuring tube wall temperatures inside a radiating firebox and proposed a method for correcting for reflected energy. He assumed that the effects of end walls in a firebox could be ignored. This is probably true for tubes toward the center of the firebox but is questionable for those tubes at or near the end walls. He also ignored gas radiation, a more serious problem.

In 1978, Grandfield [3] proposed a method for reducing error in radiant tube temperature measurement using radiation pyrometry that offered some possibility of success. His method considered reflected energy and gas radiation and involved correction equations that could be readily presented in graphical form for ease of use.

TRAINING WORKSHOP

At this point in the investigation, it became obvious that a consultant was needed. Considering the leverage accurate temperature measurements exercised on plant operating economics and the lack of expertise in this area in both organizations, Monsanto and Conoco managements agreed to bringing in an outside consultant.

The consultant services of Mr. S. D. Grandfield (now deceased) were arranged to provide a three-day workshop on radiant tube wall temperature sensing. Mr. Grandfield had 20 years experience in the manufacture and use of electro-optical components, instruments, and controls. He held patents relating to non-contact temperature measurement systems and had authored manuals and articles and conducted workshops on industrial temperature measurement and control.

The workshop consisted of two days of formal training in the theory and practice of temperature measurement and one day of field training. The first two days of the workshop were held at a site away from the plant and approximately 20 engineers (manufacturing and technical) attended. Major topics discussed were as follows:

• Radiant tube temperature sensing

• Radiation physics

- Radiation from tubes
- The Grandfield Associates correction method: sample calculations
- Non-contact sensors: methods for specifying and purchasing
- Thermal imager strategies
- Calibration of non-contact sensors
- Avoiding problems with non-contact sensors
- Optical aids to radiant tube viewing
- Tube temperature management and its environmental implications
- Contact sensors

PYROMETER EVALUATION AND SELECTION

The final day of the workshop was spent making field measurements on an operating furnace in our old ethylene plant. Operating in teams of two, each person "shot" the same spots on a tube near its outlet. Each team shot two spots on the tube and the radiant refractory wall facing each of the tube observation spots. Four different infrared pyrometers of similar characteristics were employed.

One of the pyrometer's readings drifted with prolonged, almost continuous use. This problem was reported to the manufacturer who subsequently corrected the problem. Data for each team for each pyrometer were analyzed for average reading, scatter of reading, and ranges of observations. Standard error of measurement for each pyrometer was determined. Each team commented on the ease of use of each pyrometer and their personal preference; these comments were analyzed.

Accuracy of measurements was determined using tube wall temperature values calculated using the mathematical model of the furnace. The spots at which the observations were made were carefully located with respect to radiant coil return bends or other fittings so their linear location along the coil was accurately known.

The mathematical model of the furnace is a comprehensive design model. It describes the firebox side of the furnace from fuel and air in to stack flue gas out; the process side is also described from the process inlet to the convection section and also the cracked gas out of the radiant coil. Up to five utility fluids can be handled in the convection section. The following are calculated: preheat of feed water to the boiler, superheat of steam, and heat savings obtained by preheating combustion air. Detailed mechanical descriptions are required of the radiant section coils, the process convection section coils, the utility service convection section coils, the radiant firebox, and the convection section box. Fuel composition, excess air, air humidity, compositions or characterizations of the usual feed hydrocarbons, process steam, utility streams, and fuel charge rates are inputs. Flow and thermal properties such as heat capacities, densities, viscosities and thermal conductivities along with kinetic parameters describing the cracking reactions are also inputs.

The program is designed such that any input or calculated data can be printed out as desired. Up to five dependent boundary conditions can be specified. In order to do that, an independent boundary condition is designated for each dependent condition. Typically, the radiant coil outlet pressure, the radiant coil outlet temperature, and an overall heat balance error are designated as dependent conditions; independent boundary conditions are designated as the convection section inlet pressure, the stack flue gas outlet temperature, and the fuel firing rate.

Radiant section information normally printed out are bulk process gas temperature profile, tube wall temperature profiles, hydrocarbon partial and total pressure profiles, heat flux profile, heat duty profile, composition profile, fuel rate, firebox cross-over flue gas temperature profile.

As tube wall temperatures were being observed, operating conditions in the furnace were documented, including measurement of the fuel rate and composition, the firebox cross-over flue gas temperature using a quick response thermocouple, the excess air at the firebox cross-over and at the stack, and ambient air humidity. These measurements were needed to obtain accurate data for flue gas flowrate, heat release, and air leakage. Feed to the furnace at the time was sampled and characterized according to the requirements of the furnace model.

Furnace operating conditions at the time were simulated using the model. Radiant coil outlet temperature and pressure were forced to match observed data by iteration. Radiant coil inlet temperature and radiant coil inlet pressure matched the observed data within 5.6°C (10°F) and 6895N/m² (1 psi) respectively. Calculated stack flue gas outlet temperature was within 8.3°C (15°F) of the observed values, while firebox cross-over flue gas temperature was within 13.9°C (25°F) of the observed value. Calculated fuel consumption was within 1%. This match of the model with actual operation was expected since similar checks in the past had given comparable results. The quality of this match, along with previous results, gave us confidence in the calculated values for the tube wall temperatures for the positions at which our pyrometer measurements had been made. The calculated values were used as the true value for tube wall temperatures and as the basis for accuracy determination of the pyrometers.

Based on information obtained during the workshop and the field checks, a tentative pyrometer selection was made. The manufacturer of that instrument was contacted to determine turnaround time (in their shop) for repair, and the availability, delivery, and rental of "loaner" pyrometers if such were ever needed. (Since the plant has been on-stream, one occasion has arisen when a rental pyrometer was needed. It was in the plant less than 24 hours after the manufacturer was contacted.) The final decision on the pyrometer to specify was then made. Three pyrometers were ordered: one for active plant use, one in reserve for plant use, and one for recycling for repair and/or calibration checking.

The infrared pyrometers selected were hand-held battery-operated (HHBO) units with a narrow band filter (3.5-4.1 micrometers) to eliminate interference from radiating carbon dioxide and water vapor in the combustion gases. Since the fuel used was sulfur-free, there was no need to worry about sulfur oxides. Digital output was specified. Digital output and target could be viewed simultaneously. This has proven to be a real convenience considering the number of observations that have been made.

Field of view was 3.8 cm (1.5 in) at 3.05 m (10 ft), 6.35 cm (1.5 in) at 9.1 m (30 ft) and 10.2 cm (4 in) at 15.2 m (50 ft). Tube temperature measurements were to be made from a distance of approximately 1.5-2 m (5-7 ft), thus the spot size on the tube was approximately 3-3.5 cm (1.2-2.5 in), small enough that even the smallest diameter tube could be observed without "seeing" the refractory wall behind it. Field of view was also important in selecting the proper size for the sight tube on the calibration blackbody furnace. The pyrometers also incorporated a "peak hold" circuit. This feature has, however, been of no use because of the nature of observations made. Working distance of the pyrometers are 2.5 cm (6 in) to infinity. Response time is 0.75 s.

PYROMETER CALIBRATION

With the amount of field handling the pyrometers would have, even though the observers were trained to treat them as "camera" quality equipment, it was important to have an in-plant capability for checking and maintaining the calibration of the instruments. Hence, a blackbody furnace and a secondary standard infrared thermometer were purchased.

The blackbody furnace obtained was a Land Type MT with a 320 mm (12.15 in) diameter sphere internally coated with a rough silicon carbide coating. It includes a 65 mm (2.56 in) ID Mullite sight tube, a size compatible with the field of view of the pyrometers. The furnace is electrically heated (Kanthal Al elements) using a single-phase, 200-240 V, 13 A, 50-60 Hz supply. Furnace temperature can be controlled anywhere in the range of 500–1150°C (932–2102°F) using a self-contained temperature controller with a platinum/platinum-13% rhodium thermocouple. Actual blackbody temperature is sensed using two platinum-6% rhodium/platinum-30% rhodium reference thermocouples. The reference thermocouples enter the furnace from the top at an angle to each other. They are used for calibration measurements. Use of these thermocouples for blackbody temperatures (average of the two) resulted in a calibration transfer accuracy on 12°C (21.6°F).

To improve calibration transfer accuracy to 4°C (7.2°F), a certified National Bureau of Standards traceable standard infrared thermometer was also purchased. It was mounted in the furnace sight tube using a special water-cooled adaptor.

The blackbody furnace is located in our Central Instrument Shop. Maintenance of the furnace, the secondary standard thermometer, and all field pyrometers calibrations and maintenance is done by instrument technicians specially trained on the equipment and procedures. Maintenance beyond their capability is coordinated by them with the instrument manufacturers.

Calibrations are normally checked at 650°C (1200°F) and 1038°C (1900°F). About 12-18 h time is requied to change the blackbody furnace from one temperature to another. It would be advantageous to have a second blackbody furnace, perhaps homemade, at the lower temperature. Accuracy of the cheaper furnace could be checked using the better one.

OPERATING PERSONNEL TRAINING

All technical personnel involved in operation of the plant attended Grandfield's workshop. Shift personnel making the routine observations of tube wall temperatures attended a four-hour training session. Training emphasis was an operation and care of the pyrometers, need for accurate measurements, proper and improper techniques for obtaining tube wall temperature data, proper use of data forms, and Grandfield Associates' tube wall temperature correction method. Special attention was called to problems caused by vignetting from sighting too close to the refractory edges of peephole doors. To protect the instruments from overheating because of thermal radiation exposure, one hand was to be placed in front of the instrument near, but not blocking the view of, the optics. If it was too hot for the observer's hand, it was too hot for the instrument, and the observer must back further away from the peephole to make the observation.

OPTICAL AIDS TO RADIANT TUBE VIEWING

To use infrared pyrometers, the instruments should have clear, unobstructed views of the target spots on the radiant tubes. Our furnaces were designed with this in mind. Except for the bottom 1.5 m (5 ft) every tube is clearly visible from at least two peepholes on each side of the firebox. The bottom of each tube is visible from only one peephole on each side. All return bends are visible from both sides.

Since open doors on peepholes are prime sources of unwanted air infiltration into the firebox, care should be taken to open as few as possible while providing satisfactory viewing of the tubes, both visually and with pyrometers. Peephole door design should include consideration of ample sealing against air infiltration when the door is closed.

In older furnaces, or in pressurized reactors, a clear unobstructed view of the target may not be available. It is possible to view an otherwise hard-to-see target spot using a mirror or periscope in conjunction with an infrared pyrometer. The front surfaces of the targets should be reflecting metals or metal coatings. Thin gold coatings on stainless steel substrates are one of the best, having a reflectance of unity at 4 micrometers. Materials for mirrors and their holders should be stable after some exposure in the field environment.

Vignetting can be a problem when using mirrors or periscopes. Edges of the peepholes, mirrors, or periscopes may obscure the edges of the energy "bundle" entering the lens of the pyrometer. With proper safeguards, any possible furnace opening can be used to view inaccessible tubes. A thermowell, draft connection, or oxygen sampling probe can be removed temporarily, for example. If mirrors or periscopes are used the reflecting surfaces must be kept clean and free of grease and dust.

Since there will almost always be an energy loss at a reflecting surface (10-20% per reflecting surface as a "rule of thumb") calibration of mirrors and periscopes may be necessary. This can be done by sighting the pyrometer on a stable hot object (a blackbody furnace, for example) with and without the reflector(s) in the optical path.

For pressurized reactors, targets can be viewed through windows having high transmissivity, both in the visible and pyrometer sensing wavelengths. Windows must be able to withstand exposure to their environment for satisfactory time periods. With proper installation, ordinary window glass may be used with pyrometers using the 2.1-2.5 micrometer sensing band. Calcium fluoride or synthetic sapphire may be used with 3.5-4.1 micrometer wavelength pyrometers. Calibration of windows will also be necessary since there is always some loss of rays passing through them (about 10%). Windows can be calibrated by measuring a hot target through the atmosphere with and without a window in the optical path.

DATA OBSERVATION AND RECORDS

During the first three months of our plant operation, tube wall temperature data were taken daily on all radiant coils in each furnace. Even though time consuming, it was felt necessary to emphasize the importance of tube wall temperatures, provide training and education, and to provide historical data. Information obtained during this period indicted that such detailed surveys were not necessary for routine plant operation. Full coil surveys are, however, still made occasionally, usually in connection with furnace tests or burner adjustments. If, during a routine survey, visual inspection of the coils indicates problems with hot spots or abnormal wall temperature patterns, full surveys are made. They are also used to monitor relative firing distribution between floor burners and wall burners, partially plugged wall burners, and uneven distribution of combustion air between burners.

Two different procedures are now used for routine monitoring of tube wall temperatures: a "total" survey covering the last four tubes in each coil and a "partial" survey covering only the last two tubes in each coil. The decision for a "total" or a "partial" is made by unit supervision after considering (1) how long a furnace has been on-line since decoking, (2) the feedstock and operating conditions of the furnace, (3) production demands, and (4) furnace availability at the time and in the short range future. Data are recorded on the forms shown in Figures 1 and 2. Process data, taken while the tube wall temperature surveys are being made, are recorded on a separate form and attached to the pyrometer survey sheets.

Tube wall temperatures are shot from both sides of the firebox. As measurements are made from one side, refractory wall temperatures on the opposite side are observed and recorded. Thus when a survey is completed, both tube wall temperatures and refractory temperatures directly opposite the tube spots have been measured. These refractory temperatures are recorded

directly on the forms at the approximate location relative to the coil, tube and observation peephole. For convenience in the field, refractory temperature for one side of the tubes are recorded on the form for the opposite side of the tubes.

Refractory temperatures are used to correct the observed tube wall temperatures. If localized hot spots are noticed, their temperatures are recorded on the form and circled. An arrow is then drawn from the circled data to the location of the hot spot on the appropriate tube relative to the peephole level from which the observation was made.

Our present temperature survey program requires "total" surveys on two furnaces per day and "partial" surveys on the remaining furnaces once per day. Full coil surveys in which all tubes in a furnace are checked may be done at any time needed. Since these surveys are time-consuming, tedious, and sometimes unpleasant, the work load is shared by all operating shifts.

PYROMETER ERRORS AND CORRECTIONS

Non-contact temperature measurements are open to several sources of error. The tubes are viewed through hot combustion gases. Carbon dioxide, water vapor, and other triatomic flue gas components are excellent sources of radiation that will be "seen" by a pyrometer unless the pyrometer is filtered to "look-through" windows in the gases infrared spectra. Our pyrometers are filtered to view through the 3.5-4.1 micrometer window in the carbon dioxide/water vapor spectra. Our fuel contains no sulfur so this window is satisfactory for us.

Another source of error concerns the relative quantities of radiation actually emitted by the hot tubes as compared to that reflected by the tubes from the hotter wall enclosures. An infrared pyrometer is a total energy device that responds to the total energy received by the sensor, regardless of the source of that energy. Readings for this error are corrected using a method developed by Grandfield (3). Grandfield's method used Planck's equation and a sum-of-radiance equation.

Planck's equation is:

$$W_\lambda = C_1\lambda^{-5}\,[\exp(C_2/\lambda T)-1]^{-1} \qquad (1)$$

with,

W_λ = total radiance, W/ cm^2–cm,

C_1 = $2\pi c^2 h$, W–cm^2,

λ = wavelength, cm

C_2 = hc/k, cm-K

c = velocity of light, 2.99793×10^{10} cm/sec,

h = Planck's constant, 6.6252×10^{34} W–sec^2,

k = Boltzmann's constant, 1.380043×10^{23} W–sec/K

The sum-of-radiance equation is:

$$W_{\lambda,t} = rW_{\lambda,r} + \epsilon W_{\lambda,e} \qquad (2)$$

with,

$W_{\lambda,t}$ = the total radiance from the tube available to the pyrometer

$W_{\lambda,r}$ = ambient radiance available for reflection by the tube from the hotter enclosure

$rW_{\lambda,r}$ = that portion of the ambient radiance reflected by a tube with a reflectivity of r

$\epsilon W_{\lambda,e}$ = the radiance emitted by the tube because of its temperature and emissivity

For a given value of λ, Equation (1) can be reduced to:

$$W_\lambda = k_2\,[\exp(k_2/T) - 1]^{-1} \qquad (3)$$

with,

$k_1 = C_1\lambda^{-5}$, W cm^{-2}

$k_2 = C_1\lambda^{-1}$, K

T = temperature, K

Equation (3) can be rearranged and solved to T to yield:

$$T = k_2\,[\ln(k_1/W_\lambda) + 1]^{-1} \qquad (4)$$

For our pyrometers the central sensing wavelength is 3.8 micrometers (3.8×10^{-4} cm), thus:

$$k_1 = C_1\lambda^{-5} = 2\pi c^2 h\,\lambda^{-5} =$$

$$[2\,(3.14159)\,(1.99793 \times 10^{10})^2\,(6.6252 \times 10^{-34})\,/$$

$$(3.8 \times 10^{-4})^5 = 472176/89\ \text{W cm}^{-2}]$$

$$k_2 = Pc_1\lambda^{-1} = hc\,\lambda^{-1}\,k^{-1} =$$

$$[(6.6252 \times 10^{-34})\,(2.99793 \times 10^{-10})]\,/$$

$$[(1.38042 \times 10^{-23)}$$

$$(3.8 \times 10^{-4})] = 3786.4\ \text{K}$$

Grandfield's correction is determined as follows:

1. With pyrometer emissivity control set at unity,

 a. Scan the hot refractory surface facing the plane of the tube row to be measured. An approximate average value is determined and then converted to K. This is T_R.

 b. With the same pyrometer, the temperature is measured at the points of interest on the tube surface. The hottest point on the tube circumference (on an internally clean tube) is the point with the best view of the hot furnace enclosure and the poorest view of the adjacent tubes and/or tube supports. The hottest temperature is converted to K and called T_A, the apparent tube temperature.

2. The total energy affecting the pyrometer at T_A is calculated using Equation (3).

3. The reflected energy portion of the total energy is calculated using T_R, the tube reflectivity, and Equation (3).

4. The energy corresponding to the true tube temperature, T_T, is calculated by subtraction using Equation (2), tube emissivity, and the reflected energy component.

5. The true tube temperature, T_T, is calculated using Equation (4). Convert to °C or °F.

A sample calculation using Grandfield's method and the following assumed data is shown below.

Steps 1a and 1b above:

Data has been collected with the following results,
  Tube emissivity, $\epsilon = 0.90$
  Tube reflectivity, r $= 0.10$
  Apparent tube temperature, $T_A = 1800°F$ (982.2°C, 1255.4 K)
  Refractory temperature, $T_R = 2025°F$ (1107.2°C, 1380.4 K)

Step 2:

Calculate total energy affecting pyrometer,

$$W_{\lambda, t} = 472176.89[\exp(3.786.4/1255.4) - 1]^{-1}$$

$$= 24324.80 \ W/cm^2 - cm$$

Step 3:

Calculate reflected energy portion of total energy,

$$W_{\lambda, r} = 472176.89 \ [\exp(3786.4/1380.4) - 1]^{-1}$$

$$= 32489.76 \ W/cm^2 - cm$$

$$rW_{\lambda, r} = 0.1 \ (32489.76) = 3248.98 \ W/cm^2 - cm$$

Step 4:

$$W_{\lambda, e} = (W_{\lambda, t} - rW_{\lambda, r})/\epsilon$$

$$= (24324.80 - 3248.98) \ / \ 0.90$$

$$= 23417.58 \ W/cm^2 - cm$$

Step 5:

Calculate T_T and convert to °C and °F.

$$T_T = 3786.4\{\ln[(472176.89/23417.58) + 1]\}^{-1}$$
$$= 2340.5 \ K = 969.4°C = 1773.2°F$$

Thus, under the sample calculation conditions an apparent tube wall temperature of 1800°F (982°C) corrected becomes 2773°F (967°C). As the difference between the apparent tube temperature and the refractory surface temperature increases, the amount of correction increases. For example, with the same refractory temperature used above (2025°F, 1107.2°C) but with an apparent tube wall temperature of 1500°F (815.6°C, 1088.8 K), the true corrected temperature is 1427°F (775°C).

Knowledge of the target emissivity is a "must" for accurate pyrometer measurements. To illustrate, the same conditions are assumed as used in the sample calculation above. A tube emissivity ϵ, of 0.85 and a reflectivity r, of 0.15 are, however, used. $W_{\lambda, t}$ is still 24324.80 W/cm^2 −cm. Now, however,

$$rW_{\lambda, r} = 0.15 \ (32489.76) = 4873.46 \ W/cm^2 - cm$$

$$W_{\lambda, e} = (24324.80 - 4873.46) \ / \ 0.85$$

$$= 22883.93 \ W/cm^2 - cm$$

$$T_T = 3786.4 \ \{[\ln(472176.89/11883.93) + 1]^{-1}\}$$

$$= 1231.6 \ K = 958.5°C = 1757.3°F$$

Grandfield's method of correction involves a lot of tedious calculations. For our use the method has been customized: the proper value for tube emissivity and the wavelength of our pyrometer's spectral window has been substituted into Grandfield's equations and the specific equations programmed on a hand calculator. A graphical solution, for control room use, is shown in Figure 3, which includes an example of its use. For a refractory temperature of 2100°F (1150°C) and an observed tube wall temperature value of 1600°F (870°C), the true tube wall temperature is 1535°F (835°C). As the tube wall temperature increases for a given refractory temperture, the size of the correction decreases. As tube wall temperature approaches the refractory temperature, the correction approaches zero.

For operations, corrections are usually not applied until the observed tube wall temperature reaches 1900°F(1040°C). Refractory temperatures are, however, recorded regardless of the tube temperature, so that corrections can be made if needed for any reason.

CONCLUSIONS

A workable and effective tube wall temperature monitoring program has been developed and tested with five years of good experience. Attempts have been made to minimize problems and errors. We will continue to look for improvements in equipment, techniques, or even new approaches that might give better results.

REFERENCES

1. Bolls, W. L., "Measurement of Gas Temperatures by Means of Thermocouples", Petroleum Refiner, 2, pp. 120-129, (1948).

2. Nicholson, R., "Measurement of Tube Metal Temperatures in Radiant Wall Process Furnaces Using the Disappearing Filament Pyrometer", Journal of the Institute of Fuel, pp. 258-261, (1973).

3. Grandfield, S. D., "Method Cuts Error in Radiant Tube Temperature Sensing," The Oil and Gas Journal, pp. 68-70, (1978).

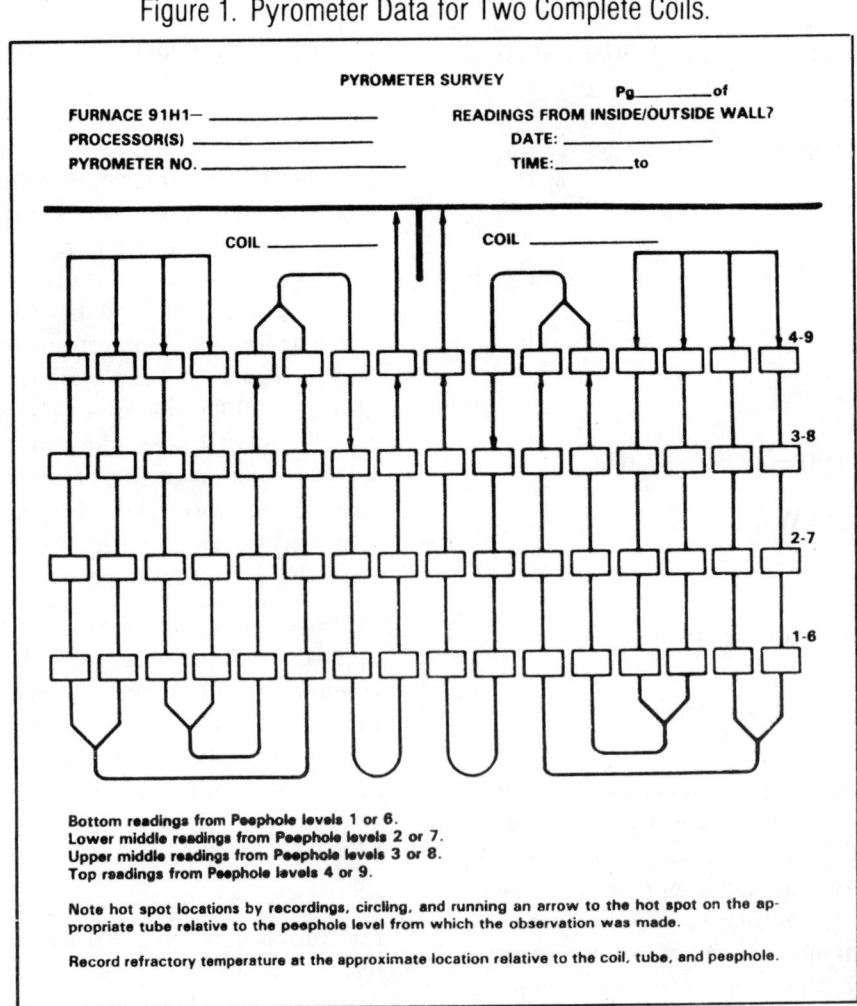

Figure 1. Pyrometer Data for Two Complete Coils.

Figure 2. Pyrometer Data for Portions of Two Coils.

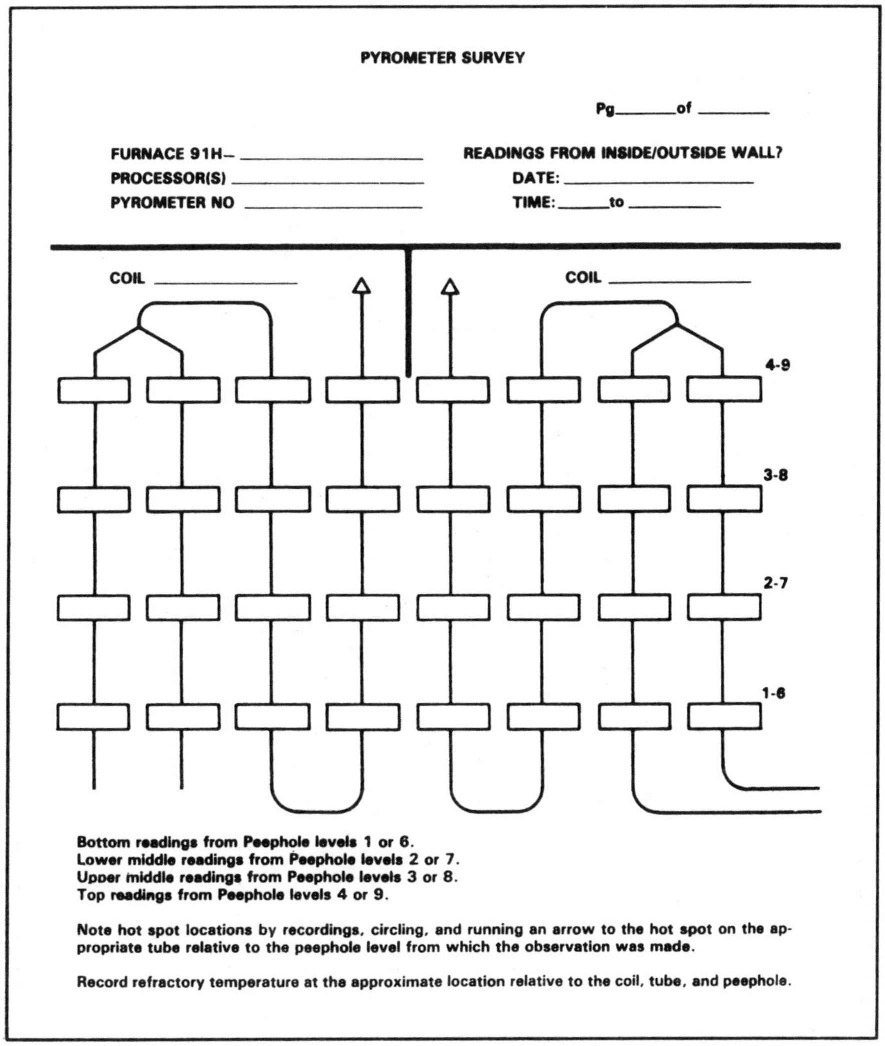

PYROMETER SURVEY

Pg_____of _____

FURNACE 91H– _____   READINGS FROM INSIDE/OUTSIDE WALL?

PROCESSOR(S) _____   DATE: _____

PYROMETER NO _____   TIME: _____to _____

COIL _____   COIL _____

4-9

3-8

2-7

1-6

Bottom readings from Peephole levels 1 or 6.
Lower middle readings from Peephole levels 2 or 7.
Upper middle readings from Peephole levels 3 or 8.
Top readings from Peephole levels 4 or 9.

Note hot spot locations by recordings, circling, and running an arrow to the hot spot on the appropriate tube relative to the peephole level from which the observation was made.

Record refractory temperature at the approximate location relative to the coil, tube, and peephole.

Figure 3. Tube Wall Temperature Correction.

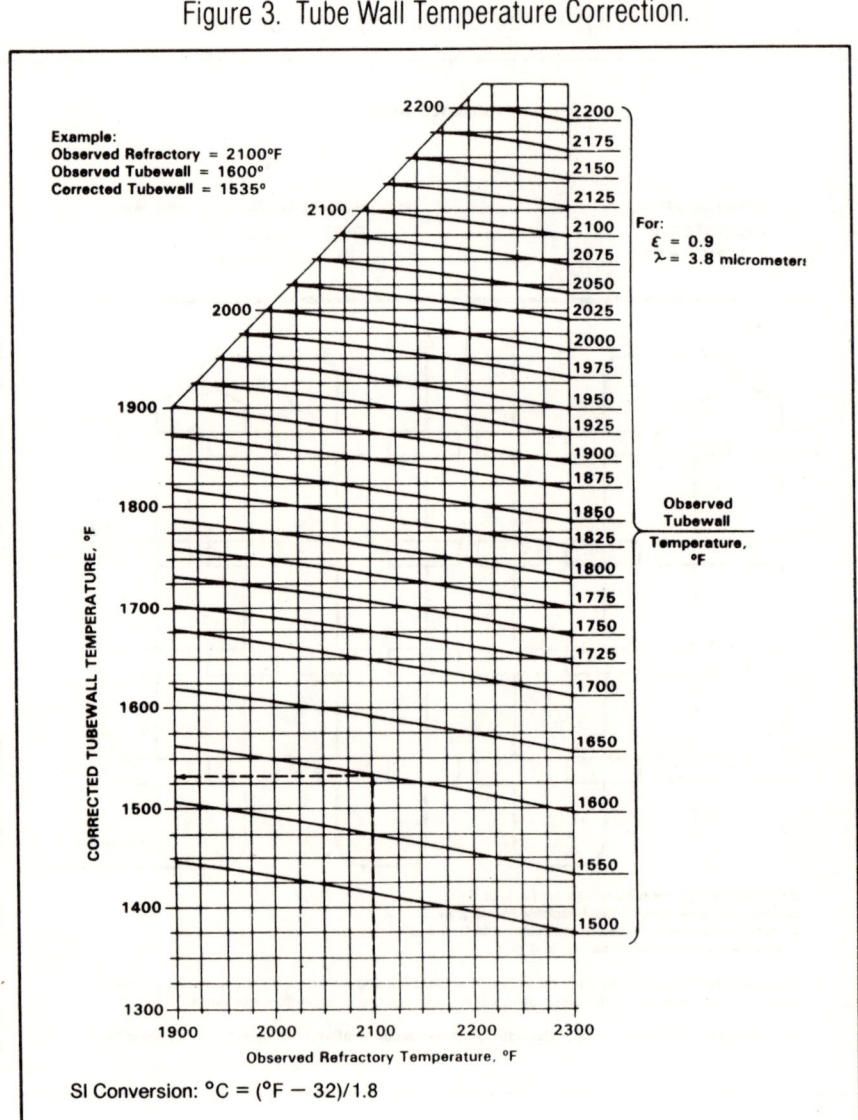

RADIOMETRIC TEMPERATURE MEASUREMENT OF FURNACE LOADS: OPTIMUM CHOICE OF OPERATING WAVEBAND

T.G.R. Beynon and R. Barber ■ Land Infrared Ltd., Sheffield, England

Correct choice of operating wavelength is of paramount importance in radiation thermometry; particularly so in thermometry on furnace loads, where radiation reflection and atmospheric emission/absorption can seriously influence the measurement. In the specific, but by no means singular, case of steel billet reheating furnaces, a strong argument can be made for a narrow waveband centered on 3.9μm; moreover, this choice has been validated in extensive trials. The rationale behind the adoption of 3.9μm for reheat furnaces is presented, and should have a direct bearing on many other furnace load problems.

INTRODUCTION

Measurement of the temperature of a furnace load is desirable in a multiplicity of industrial processes, but is always difficult, often to the point where bizarre control strategies evolve which circumvent the measurement, usually paying a price in both fuel efficiency and product quality. Whether the load is the product being heated (as in a billet furnace) or is a heat exchanger tube through which reactants are passing (as in a reformer furnace), the element of the problem is the same: namely, a comparatively cool object inaccessibly situated inside a hot combustion enclosure. Thermocouples, or other conventional contact devices, require positive thermal attachment to the object to give meaningful results and this is frequently impracticable. There is a very strong incentive to employ non-contact techniques, of which infrared radiation thermometry is by far the most highly developed for use in process plants.

In the furnace environment, radiation thermometry is complicated by several effects, most notably reflection of radiation originating from the hot enclosure and emission of radiation by the atmosphere in the sight path. Constraints imposed by plant operating practice often severely restrict the means available to suppress these problems. However, a degree of freedom which is, almost invariably, open to the infrared system designer is choice of operating wavelength. In certain cases, an astute decision here can have a surprisingly powerful effect on the magnitude of the interferences, to the extent that quite crude suppression or compensation methods become sufficient to correct the measurement to adequate precision.

A case in point is that of steel billets in billet reheating furnaces. Here, adoption of a carefully defined wavelength band near 3.9μm leads to a very simple measurement system which is, in a majority of applications, sufficiently precise to form a basis for furnace control. Most importantly, the system is sufficiently rugged and maintenance-free to be widely acceptable to operators of process plants. This paper examines the rationale behind the 3.9μm reheat furnace measurement, which should have a direct bearing on many other furnace load problems.

THE FURNACE LOAD PROBLEM

There are, in general, several obstacles to successful radiation thermometry on furnace loads:

- Background reflection: reflection of radiation, originating from the hot background, off the load into the thermometer,

- Sightpath opacity: emission/absorption/scattering of radiation by the furnace atmosphere in the sight path between the thermometer and the load,

- Load emissivity: the ubiquitous problem of uncertainty in the target surface emissivity,

- Scale: uncertainty in how the temperature of the "optical" surface relates to that of the underlying substrate, and

- Practicalities: access, instrument cooling and purging, for example.

It is perhaps worth emphasising three points: firstly, only radiation *reflecting* off the target spot (that is, the intersection of the thermometer field-of-view with the load) should affect the thermometer reading; secondly, *background* includes the furnace atmosphere; and thirdly, in industrial control situations, it is *repeatability* which is usually of prime concern, rather than absolute accuracy, of measurement.

REHEAT FURNACES

In steel billet reheating furnaces (Figures 1a and 1b), the problem is characterized by: high levels of background reflection; moderate (outside gaseous absorption bands) sightpath opacities; relatively high and well-defined emissivities; generally tolerable uncertainties due to surface scale; and very severe limitations on what is practically acceptable.

Effective furnace control[1] requires load measurement in early zones where large load-background temperature differentials occur (Figure 2). Billet-wall differences of $400^{\circ}C$ are not atypical, with the atmosphere $200^{\circ}C$ above the walls and the flame cores hotter still.

Many furnaces employ gaseous fuels and, in general, have atmospheres which are relatively free from particulates outside well-localized flames; flame luminosity is often quite low, but varies with installation and fuel type. Furnaces using oils produce, with the very best combustion arrangements, atmospheres not dissimilar to those found in gas firing; but, more generally, flames are large with very bright cores, while a "haze" of particulates is visible everywhere. In extreme cases of poor fuel atomization, a "hot fog" of burning fuel permeates the furnace. Excepting

extreme examples, thermometers can be located so as to avoid luminous flame directly in the sight path and usually sight path lengths can be limited to about two meters.

Load emissivity is generally high, stable, predictable, and insensitive to viewing angles up to 50° or so off the surface normal. Wavelength dependence is slight: Table 1 gives some values based on laboratory measurements. Reflectance is basically diffuse, but with some directionality.

The measurement precision (repeatability) needed as a basis for control in preheat and heat zones is only of the order of $\pm 25^{\circ}C$, with twice this figure sufficient to prevent some of the grosser malpractices currently perpetrated. At this level, the influence of surface scale is not intractable.

The critical practical constraint is plant operators' vehement resistance to any structures, particularly water cooled, inside the furnace.

CHOICE OF OPERATING WAVELENGTH

Gas Opacity

Sightpath emission is potentially the most catastrophic interference to measurement. The furnace atmosphere is a good deal hotter than the billet and, if significantly opaque, much brighter. A prerequisite on operating waveband is that it should lie in a good atmospheric window where at least the furnace *gases* are highly transparent.

Table 1: Measured Emissivities for Mild Steel Samples.

Sample	Spectral Emissivity		
	$1\ \mu m$	$4\ \mu m$	$8\ \mu m$
37	0.90	0.84	0.81
A1020	0.93	0.84	0.83
1	0.90	0.86	0.86
2	0.89	0.86	0.87
04	0.90	0.82	0.79
A1024	0.90	0.83	0.81

The principal absorbers in the near infrared are water vapour and carbon dioxide; available data[2,3] allow computation of sight path opacities using statistical models to represent the band structures; Figure 3 shows some representative results. There is little gas opacity at wavelengths less than 1 μm and windows centered on 1.6, 2.2 and 3.9 μm.

Background Reflection

Working in a window and assuming, for the moment, that there are no particles in the sight path, the radiation signal received by a thermometer viewing a billet can be written:

$$S = \epsilon_s f(T_s) + (1 - \epsilon_s) f(\tilde{T}) \qquad (1)$$

where T_s is the billet surface temperature at the target spot; ϵ_s is the billet emissivity; $\tilde{T}$ is an effective background temperature; $f(T)$ is, apart from a constant multiplier, the Planck function (Appendix) evaluated at the thermometer operating wavelength. The first and second terms in the equation represent, respectively, emission and reflection from the target spot. The reflected component is spurious, i.e., unrelated to billet temperature, and must be suppressed or compensated.

Equation (1) is really only formalism - all the complexity of the situation lies hidden in $\tilde{T}$ - but allows some useful insights. It is particularly interesting to calculate the ratio of reflected to emitted components as a function of operating wavelength for a plausible case. Figure 4 shows results for $T_s = 800^{\circ}C$, $\epsilon_s = 0.8$, and various $\tilde{T}$. The figure shows rather clearly why silicon cell thermometers (operating wavelength 1 μm) have not worked well in the reheat application. More generally, there is a strong case for working at as long a wavelength as possible, certainly longer than 3 μm or so. The point is not simply that reflection is *reduced* at longer wavelengths, but rather that a less precise estimate of $\tilde{T}$ is necessary to effect a satisfactory correction to the thermometer signal.

Operation at 3.9 μm

The combined requirement of working at as long a wavelength as possible, yet in a good window, uniquely specifies 3.9 μm as the operating wavelength most appropriate to the

reheat application. Detailed analysis of the window residual opacity and sensitivities of available detectors lead to commissioning of the spectral-defining filter shown in Figure 5. This waveband is accessible with several detectors/optical materials well suited to plant use.

BACKGROUND COMPENSATION

Even working at 3.9 μm, background reflection is still generally too high to be tolerable without some form of suppression or compensation. Clearly, from Equation (1), if an estimate can be made for $\tilde{T}$, then a correction can be applied to the thermometer signal. It is interesting to require how precise an estimate of $\tilde{T}$ is, in fact, needed.

Accuracy Required

If an estimated value T_z were available for $\tilde{T}$ then a background correction could be applied (electronically) to the thermometer equivalent to subtracting a radiation signal $(1 - \epsilon_s)f(T_z)$. The thermometer would then indicate a temperature:

$$T_o = f^{-1}\left\{ \frac{1}{\epsilon_s}\left[\epsilon_s f(T_s) + (1 - \epsilon_s)[f(\tilde{T}) \qquad (2) \right.\right.$$
$$\left.\left. - f(T_z)] \right]\right\}$$

where f^{-1} signifies "linearization" - i.e., inversion of the calibration function and the thermometer's *emissivity control* is assumed correctly set to ϵ_s. Evidently the residual error $\Delta T = T_o - T_s$ depends on how closely T_z approximates $\tilde{T}$.

At 3.9 μm, the Planck function is sufficiently linear with temperature over the range of interest to allow (roughly):

$$\Delta T = T_o - T_s \approx \frac{1 - \epsilon_s}{\epsilon_s}\left[\tilde{T} - T_z \right] \qquad (3)$$

or, taking a practical emissivity value $\epsilon_s = 0.8$ for oxidized steel:

$$\Delta T \approx 0.25\left[\tilde{T} - T_z \right] \qquad (4)$$

This implies that, to correct the thermometer to the desired $25^{\circ}C$ precision, $\tilde{T}$ need only be estimated to within about $100^{\circ}C$. Clearly, at 3.9 μm, quite crude methods of estimating the background effect may become realistic.

Definition of $\tilde{T}$

$\tilde{T}$ may be defined as follows:

$$f(\tilde{T}) = \int_h r(\theta,\phi)f[T(\theta,\phi)]d\omega \qquad (5)$$

where $T(\theta,\phi)$ is the brightness temperature distribution within the furnace as viewed from the target spot on the load. It is just what would be measured if a 3.9 μm thermometer could be placed at the target spot and scanned around the furnace. $r(\theta,\phi)$ is a weighting factor which accounts for directionality in the reflection. It can, perhaps, best be visualized as follows: imagine that a beam of radiation (at the thermometer operating wavelength) emerges from the thermometer and reflects off the target spot; the angular distribution of the reflected radiation is precisely $r(\theta,\phi)$. Clearly the form depends on the degree of roughness (non-specularity) of the billet surface. The integral in Equation (5) is taken over the full hemisphere visible from the target spot.

Two-Part Model

It is usually reasonable to represent $T(\theta,\phi)$ by just two temperatures, T_w and T_f, corresponding to wall and flame brightness temperatures, respectively, giving

$$f(\tilde{T}) = f(T_w) + \psi\left[f(T_f) - f(T_w)\right] \qquad (6)$$

or, roughly, at 3.9 μm:

$$\tilde{T} \approx T_w + \psi[T_f - T_w] \qquad (7)$$

Here ψ is essentially the fraction of the "background hemisphere" filled by flame, but weighted by $r(\theta,\phi)$. It is, of course, not accurately calculable, but useful limits can sometimes be set or sensible measures taken to limit the value.

Particle haze, external to luminous flame, was generally insufficient to grossly influence background reflection, leaving Equation (7) valid, and a system error, taking $\tilde{T}$ as the wall temperature:

$$\Delta T \approx 0.25\,\psi\left[T_f - T_w\right] \qquad (8)$$

With $T_f - T_w$ typically $400^{\circ}C$, this is tolerable provided measures can be taken to restrict the value of ψ.

In furnaces with excellent combustion, there is little difficulty since the flames are largely confined within burner ports. More generally, one may attempt to minimize ψ by careful thermometer location. For example, the vertical viewing thermometers in Figure 1a achieve $\psi < 0.5$ on symmetry grounds alone. On the basis that $r(\theta,\phi)$ may be approximated by a specular plus a pure diffuse component, the side-viewing thermometers would benefit from fitting at a compound angle, sighting into the "nose", so as to avoid flame both in specular reflection and directly above the target spot.

RESIDUAL OPACITY

The carbon dioxide and water vapour absorption coefficients used in defining the 3.9 μm waveband are empirical values which should be reliable; the choice of band model is unimportant in a window since all models reduce to Beer's law in the limit of low opacity. However there remains the question of particles, which will emit/absorb at all wavelengths, and (conceivably) of unidentified gaseous absorbers.

For a less than perfectly transparent sight path, Equation (1) is modified to

$$S = \alpha f(T^*) + (1 - \alpha)\left[\epsilon_s f(T_s) + (1 - \epsilon_s)f(\tilde{T})\right] \quad (9)$$

where α is a sightpath *opacity* and T^* is an effective temperature which, at least in the case of gaseous opacity, will be the gas temperature. The first term above represents emission from the sight path; the second term the previous radiation signal, now attenuated. Again this is just formalism; the complexity of the situation is hidden in T^*, which will differ from the gas temperature if the principal mechanisms are particle emission or scattering. For a given value of α, the influence on the measurement is much reduced at longer wavelengths (like background reflection) simply due to the form of the Planck function.

Residual Gas Opacity

Calculated values of residual opacity in the 3.9 μm window (as defined in Figure 5) due to carbon dioxide and water vapor are given in Table 2. They are satisfactorily low at the temperatures and path lengths encountered in reheat furnaces. The carbon dioxide data is not as complete as one might wish, but the limits given in Table 2 should be reliable.

Particle Opacity

Particle opacity is practically impossible to address theoretically except to anticipate[4] that α will either decrease or remain invariant with wavelength, depending on the particle size distribution. Two types of trial were therefore made: firstly, a direct test at 3.9 μm under experimental conditions; secondly a survey, at silicon cell wavelengths, of a large number of production furnaces.

MEFOS Experiment

An experiment was made using the experimental walking-beam furnace at Metallurgiska Forkningsstationen (MEFOS), Lulea, Sweden in August 1982. A 3.9 μm thermometer was aimed across the middle zone of the (3-zone) furnace and out through an aperture in the far wall. The sight path in the hot atmosphere was 2.2 m. A large, cold, non-reflective plate was placed some meters behind the aperture and thermometer readings were taken at various fuel flow rates, firing both light

Table 2: Waveband opacities at 3.9 μm in 3 meter sight path at 1 atm. pressure, 10% H_2O, 10% CO_2. (Calculated values.)

Temp.	$\alpha(H_2O)$	$\alpha(CO_2)$
300	1.3×10^{-3}	$< 10^{-2}$
600	1.5×10^{-3}	$< 10^{-2}$
1000	3.1×10^{-3}	$< 10^{-2}$
1500	9.4×10^{-3}	$< 10^{-2}$
2000	2.8×10^{-2}	$(< 10^{-2})$
2500	5.2×10^{-2}	$(< 10^{-2})$

and heavy oils. For comparison, silicon cell and 8-14 μm thermometers were also tested.

The experiment directly measured the emission $\alpha f(T^*)$ in 2.2 m of sightpath; this may be taken as an overestimate of the signal error in a billet measurement since the attenuation effect partly compensates there. It is convenient to express the data as equivalent errors in indicated temperature (Figure 8). Adopting plausible limits for T^* allows limits to be set on α (Table 3).

The experiment was a very severe test of the thermometers used. Any "out-of-field" sensitivity, due to imperfections in the thermometer optics, led to a spuriously high reading. Furthermore, small-angle scattering in the case of billet measurements, acted to diffuse the target spot. The data in Figure 7 must therefore be regarded as upper limits.

The 3.9 μm errors, while not negligible, were satisfactorily low; the silicon cell instrument suffered from the "Planck function" effect at large atmosphere-target differentials; the 8-14 μm instrument suffered catastrophic errors and saw only the gas. It is emphasized that these results were obtained under normal combustion conditions; in extreme conditions of fuel atomization failure (easily visible in the atmosphere), gross errors resulted in all three wavebands.

Gas-Fired Furnaces

In gas firing, one anticipates quite low $T_f - T_w$ differentials at 3.9 μm: firstly, the flame gases emit little at 3.9 μm; secondly, the flames are usually optically thin in the visible and the emissivity of sub-micrometer particles is much lower at 3.9 μm than at visible wavelengths[4].

A number of furnaces were surveyed using portable 3.9 μm equipment and low $(T_f - T_w)$ differentials consistently found. The implication of Equations (5) and (7) is that, in these furnaces, a wall brightness temperature forms an adequate estimate of $\tilde{T}$.

Two-Sensor Systems

This leads directly to the very simple measurement system shown in Figure 6a. The first 3.9 μm thermometer views the billet and the second views a representative portion of the furnace wall. A processor uses the reading of the

Table 3:   Sight Path Opacities Measured at M.E.F.O.S. (2.2 m sight path).

	Flow (%)	Sight Path Opacity x 10^2			Temperature Assumed ($^{\circ}$C)
		1 μm	3.9 μm	8-14 μm	
Fuel-Light Oil					
	25	0.1	0.6	12	1000
		0.02	0.4	9.3	1200
	50	0.2	0.6	11	1000
		0.03	0.4	8.8	1200
	75	0.2	0.8	27	1100
		0.04	0.6	22	1300
	100	0.4	1.2	23	1100
		0.1	0.8	27	1300
Fuel-Heavy Oil					
	25	0.1	0.7	15	1100
		0.03	0.5	12	1300
	50	0.2	0.6	17	1100
		0.06	0.4	14	1300
	75	0.6	1.3	30	1100
		0.1	0.7	24	1300
	100	0.8	1.5	37	1100
		0.2	1.1	30	1300

second thermometer to correct that of the first as indicated above.

During the surveys it was further noticed that zone thermocouples, invariably present in reheat furnaces, usually represent the local wall brightness temperature quite well - say to 50°C or so, which is adequate in view of Equation (4). This leads to an even simpler system, Figure 6b, where the background correction is derived from a thermocouple, either existing or specially fitted. Some care is obviously necessary with thermocouple location and fixture. Figure 7a shows an installed thermometer/thermocouple pair and Figure 7b shows some installed processors.

Oil-Fired Furnaces

In a similar survey of oil-fired furnaces, flame core temperatures of 1500-1600°C were recorded with 100°C difference between 1 μm and 3.9 μm measurements. This presumably reflects the influence of coked atomized droplets with large diameters. Flames look somewhat smaller at 3.9 μm than at 1 μm.

Furnace Surveys

A large number of production furnaces have been tested similarly (i.e., sighting across to an open port) using commercially available portable silicon cell thermometers with very high quality, narrow-angle, reflex optics. (Comparable 3.9 μm instruments were, and are still not, available.)

In furnaces using gaseous fuels, atmospheric opacity is consistently found to be well below the level at which serious measurement errors arise. In oil-fired furnaces the situation is very variable, but rarely prohibitive to measurement if steps can be taken to restrict sight-path lengths. This has been successfully accomplished in a number of production furnaces by using refractory sighting tubes with a gentle flow of purge air. Such tubes are not a totally unacceptable imposition to many furnace operators, provided they are positioned so as to preclude any possibility of being struck by billets and configured so as to be replaceable with the furnace running.

SYSTEM TRIALS

Brief accounts of controlled trials for temperature measurements in various plant systems are given here; more details are provided in Reference 1.

Trial at CSM (Gaseous Fuels)

This work was carried out in collaboration with Centro Sperimentale Metallurgico (CSM) on the experimental furnace at Italsider Genoa, Italy in March 1982.

Figure 9 shows the experimental arrangement. The mild steel billet was supported on a refractory plinth and instrumented with 12 thermocouples. A removable water-cooled cover was used to produce billet/furnace temperature differentials. A 3.9 μm thermometer viewed the billet from above; background temperature was taken from a second thermometer viewing the roof. We were able to burn natural gas, blast furnace gas, coke-oven gas, gas-oil and mixtures; and were able to produce very luminous, fuel-excess flames on the latural burner. Respresentative sample data are shown in Figure 10.

The reference (thermocouple) temperatures shown in Figure 10 are steel surface temperatures extrapolated from the readings of the billet thermocouples. There was little uncertainty in the extrapolation. The system temperatures are those obtained with emissivity input as 0.82, a good value for oxidized mild steel at 3.9 μm. The system is not overly sensitive to emissivity setting within the spread obtained in sample measurements on a given material.

Throughout the trial, system errors were typically $15^{\circ}C$ or less. Large errors occured only in two circumstances: firstly, on a single occasion, the water-cooled plate shocked the billet scale layer free, so that it could be seen lying in loose pieces on the surface of the billet. In this condition, $70^{\circ}C$ errors were recorded. Secondly, there was evidence of the system beginning to read high (by about $30^{\circ}C$) at the highest temperatures tested ($\approx 1200^{\circ}C$); we attribute this to loosening of the scale layer. Overall, the system consistently met or exceeded expectation, working well under every firing condition we were able to contrive.

Trials at MEFOS (Heavy Oil Fuels)

Trials were made in collaboration with Metallurgiska Forskningsstationen (MEFOS) on the experimental walking-beam furnace at Lulea, Sweden in August 1982 and May 1983. The furnace was fired on heavy oils.

The 1982 trial was an extensive study, testing not only the 3.9 μm system per se, but also, explicity and individually, the underlying theoretical assumptions. Direct measurements were made on atmospheric opacity (see earlier discussion on Figure 8); background signals were taken from both thermocouples and radiation thermometers; data was reduced, off line, for a range of emissivity values; the entire program was carried out for silicon cell, 3.9 μm, and 8-14 μm operating wavebands. Reference temperatures were measured with a surface (gold-cup) pyrometer, subsequently compared against trailing thermocouples imbedded in the billets.

The 1983 work was a straightforward trial of the 3.9 μm system against trailing thermocouples. System data was reduced on-line using a nominal emissivity value of 0.8. Production conditions were simulated by loading up to 27 billets simultaneously. Measurements were made on both mild and stainless steel.

Figure 11 shows the basic furnace arrangement used in both trials. The billet and roof thermometers and the roof thermocouple were all in the vertical target plane A-A in the middle zone. The target billet was entered cold; walked to the measurement position as quickly as possible; and then allowed to heat. The twin oil flames were directly above. The thermometer viewed the billet at approximately 45° to the surface normal. The excellent furnace control system permitted accurate fuel rates, oxygen content, etc., to be set and recorded.

1982 Results. Several runs were made, comparing the 3.9 μm system against the surface thermocouples. Figure 12 shows an example. The system's relative insensitivity to emissivity setting is apparent. The error bars correspond to $0.80 < \epsilon < 0.84$ which more than covers the scatter found in laboratory measurements on billet samples. It is also apparent that it makes little difference whether the background temperature is derived from a second radiation thermometer or a thermocouple.

1983 Results. Good data was obtained for five consecutive runs on mild steel. In each run the furnace temperature was steady at about 1250°C; the fuel rate was close to 100% at the start of measurement, falling steadily to about 50% when the flames were cut; the excess oxygen content was 2%. Billets were approximately 100 x 100 x 1500 mm. Generally seven billets were loaded, three preceding and three following the target billet; but in run 3.1, the total number was 27. The target billet was instrumented with a thermocouple 10 mm below the surface at the measurement position. System temperatures were taken from billet thermometer corrected, on-line, from roof thermocouple, with emissivity set to nominal 0.80.

Figure 13 shows composite sample data for all runs. The graph illustrates very well the highly consistent relationship observed between 3.9 μm system temperature and sub-surface steel temperature.

A single run was made with a stainless steel target billet. System versus 10 mm sub-surface thermocouple temperatures are shown in Figure 14.

USA Results.

Parallel trials were made by the Bethlehem Steel Corporation in the USA using equipment supplied by the authors' company and supported by their subsidiary Land Instruments, Inc. These included production furnace evaluations.

Bethlehem research furnace data closely corroborates that obtained at CSM and MEFOS. Mill evaluations, under a wide range of conditions, showed an overall repeatability of 19°C (standard deviation)[5]. Systems are currently installed in plants at Burns Harbor, Sparrows Point and Bethlehem. At Sparrows Point, a system has been in continuous operation for more than two years.

CONCLUSIONS

The reheat furnace application is a good example of how important choice of operating wavelength can be in radiation thermometry. More generally it exemplifies how a radiation thermometer can be tailored to suit a specific application. All too often radiation instruments are selected simply on grounds of convenience and availability, and consequently poor performance often results.

The reflection, opacity, emissivity and scale problems discussed above require consideration in every furnace load application. However, the 3.9 μm waveband adopted for reheat furnaces should not be regarded as universally appropriate. The specific factors which favor 3.9 μm are:

- Large temperature differentials
- Moderate sight paths
- Well oxidized targets

In very uniform enclosures the "Planck function" disadvantage of short wavelengths may be tolerable, allowing rather simpler instruments, based on silicon or germanium detectors, to be employed. The work above does not establish the transparency of the 3.9 μm window over very long sight paths or at very high temperatures. Bright targets are a difficult problem, usually necessitating some form of screening or on-line emissivity measurement, and may benefit from a different choice of waveband.

Studies are in progress to identify other furnace load measurements which could benefit from adoption of the 3.9 μm waveband; reformer tube furnaces are, on the face of it, an interesting possibility. However, quite disparate views seem to exist as to the relative importance of the opacity, reflection and emissivity problems in these furnaces. Assuming that these views truly represent differences of particular experience, then considerable work is needed to elucidate the different situations and define optimum approaches.

REFERENCES

1. Beynon, T.G.R. and Ridley, I., "Direct Measurement of Furnace Load Temperature Using 3.9 Micrometer Radiation", *High Temperature Technology*, Vol. 2, No. 2, May 1984.

2. "Handbook of Infrared Radiation from Combustion Gases", NASA Special Publication 3080, 1973.

3. Nettleton, D.H., "Carbon Dioxide Absorption", National Physical Laboratory, May 1980, (Commissioned by Land Infrared, Ltd).

4. Stull, V.R. and Plas G.N., *J. Opt. Soc. Am.*, Vol. 50, No. 3, 1960.

5. Peacock, G.R. and Martocci, A.P., "A Practical Method for Direct Steel Temperature Measurement in Reheat Furnaces", Association of Iron and Steel Engineers, 1984 Spring Convention, Dearborn, Michigan, April 1984.

APPENDIX

For reference, from the Planck Function, the spectral radiance is

$$B(\lambda,T) = \frac{C_1 \lambda^{-5}}{e^{C_2/\lambda T} - 1} \qquad [\text{watt-m}^{-2}\text{-sr}^{-1}]$$

with $C_1 = 1.19106 \times 10^{-16}$ watt-m^2-sr^{-1}
$C_2 = 1.43879 \times 10^{-2}$ m-K

and wavelength and temperature expressed in meters and kelvins (K), respectively. At a temperature T_o, the distribution peaks at a wavelength λ_p given by:

$$\lambda_p T_o \approx C_3$$

where $C_3 = 2.898 \times 10^{-3}$ m$-$K $\approx 3 \times 10^{-3}$ m$-$K. At reheat furnace temperatures, typically $\lambda_p \approx 2$ to 3 μm.

Shortward of λ_p, the variation with temperature is approximately exponential according to Wien's law: that is,

$$B(\lambda,T) \approx C_1 \lambda^{-5} e^{-C_2/\lambda T} \qquad (\lambda \ll \lambda_p)$$

Beyond λ_p the variation with temperature is approximately linear according to the Rayleigh-Jeans law; that is

$$B(\lambda,T) \approx \frac{C_1}{C_2} \lambda^{-4} T \qquad (\lambda \gg \lambda_p)$$

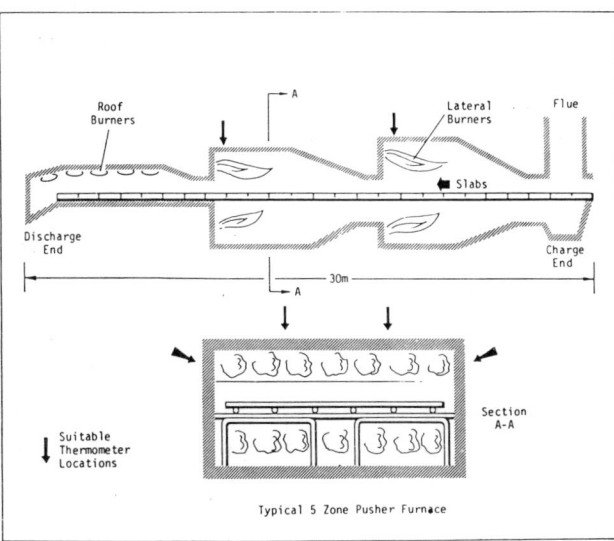

Figure 1a. Typical 5-zone pusher furnace.

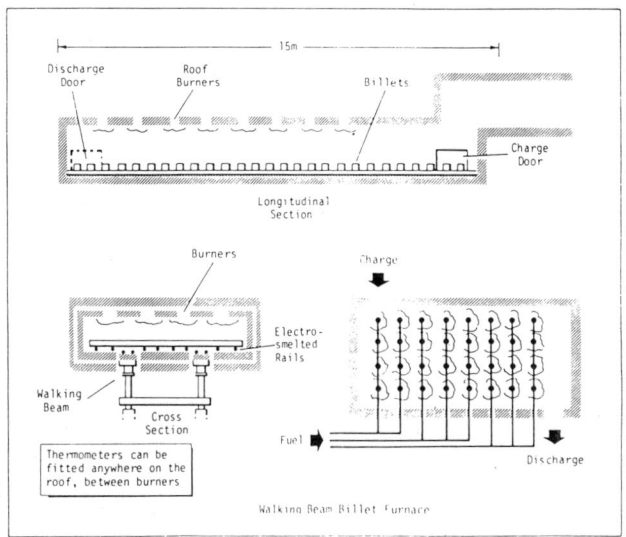

Figure 1b. Typical walking beam billet furnace.

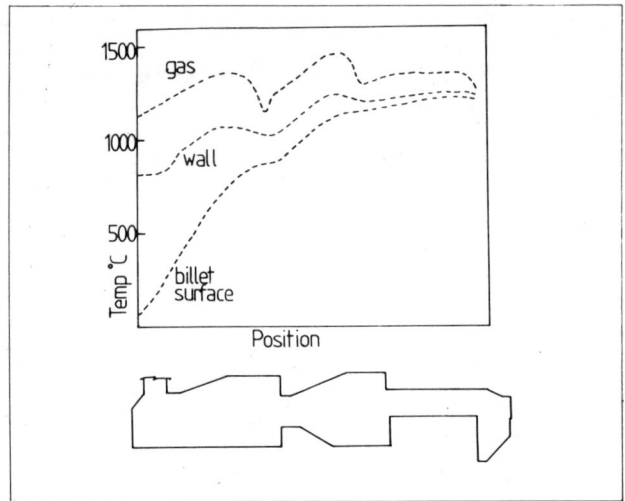

Figure 2. Typical temperatures in reheat furnaces.

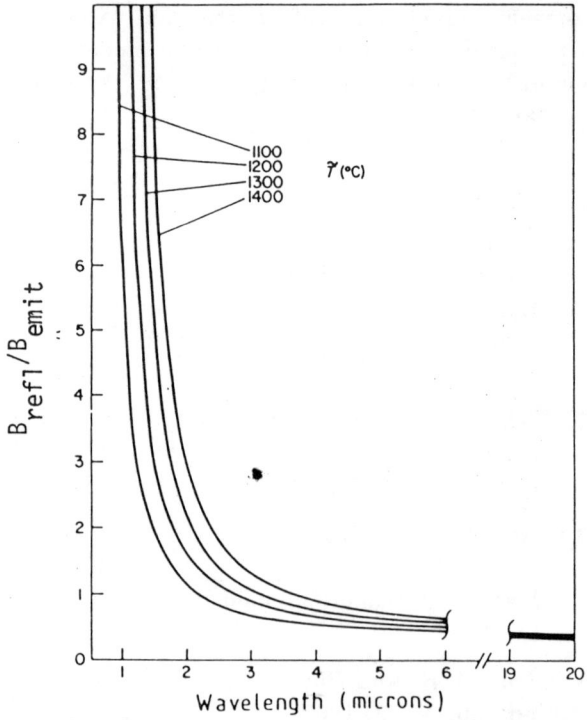

Figure 4. Ratio of reflected radiance to emitted radiance versus wavelength for 800° C billet; emissivity 0.8; various mean background temperatures.

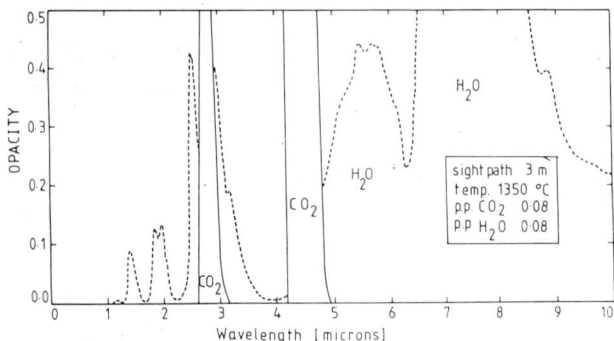

Figure 3. Opacity of typical furnace gases versus wavelength; 8% CO_2 & H_2O by partial pressure; 3 m sight path; 1350° C.

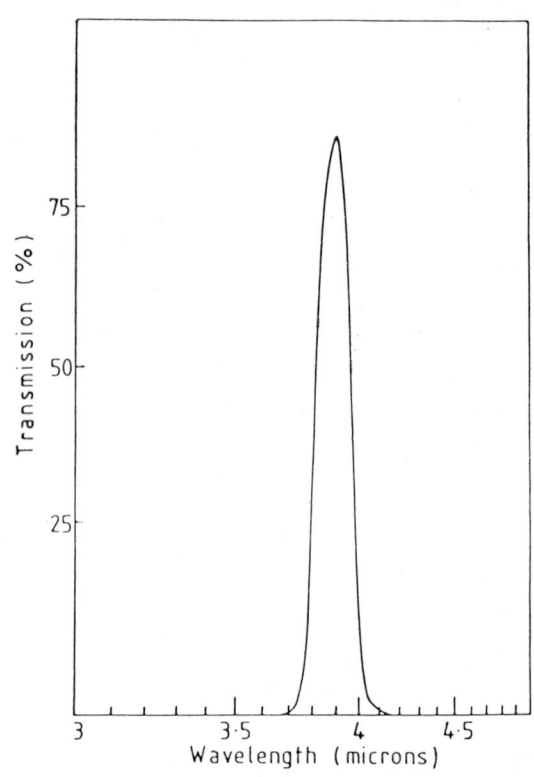

Figure 5. Thermometer operating waveband at 3.9 μm.

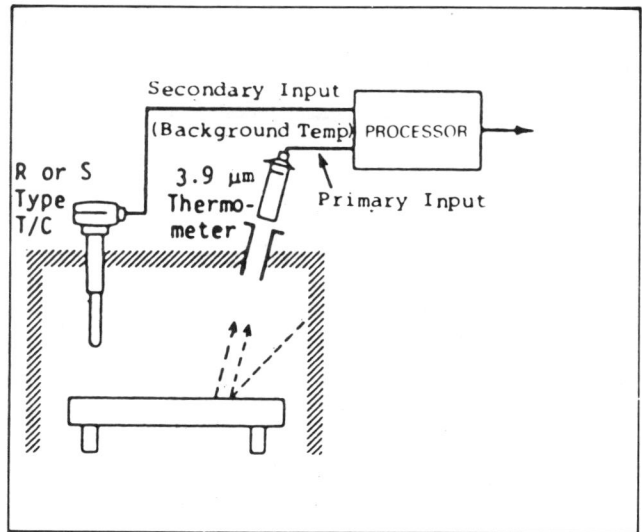

Figure 6a. Billet temperature measurement using a 3.9 μm thermometer and a zone thermocouple.

Figure 7a. 3.9 μm thermometer and thermocouple installation.

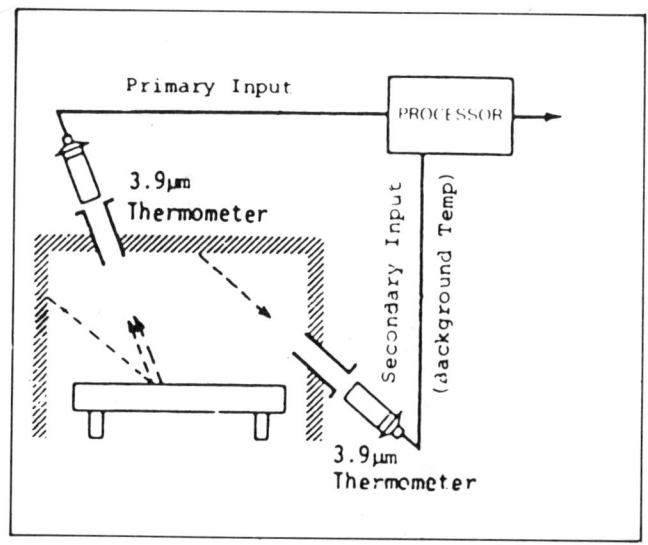

Figure 6b. Billet temperature measurement using two 3.9 μm thermometers.

Figure 7b. Installed processor units.

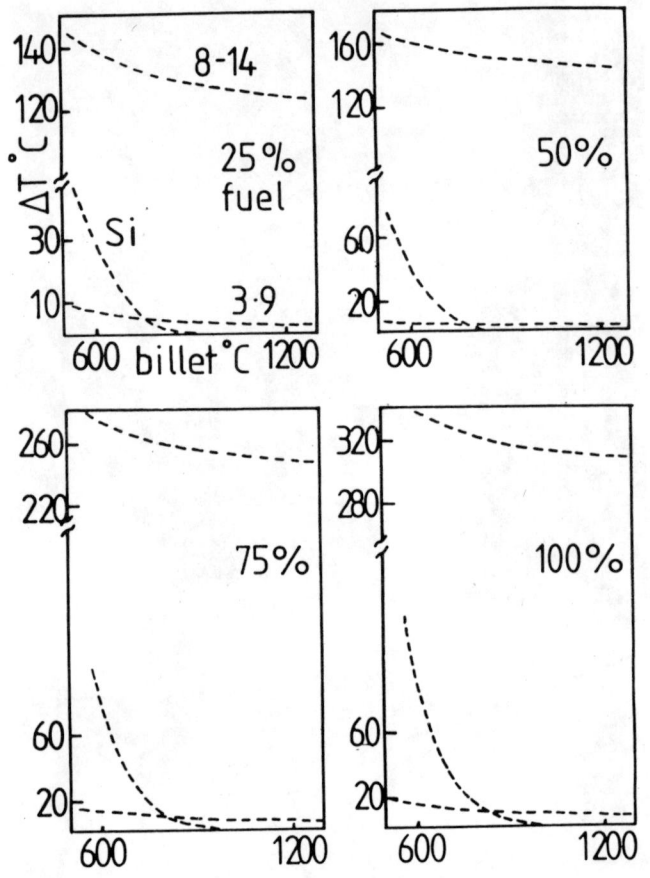

Figure 8. Measurement error, $\Delta T(°C)$, due to sightpath emission in MEFOS Furnace (heavy oil, 2.2 m).

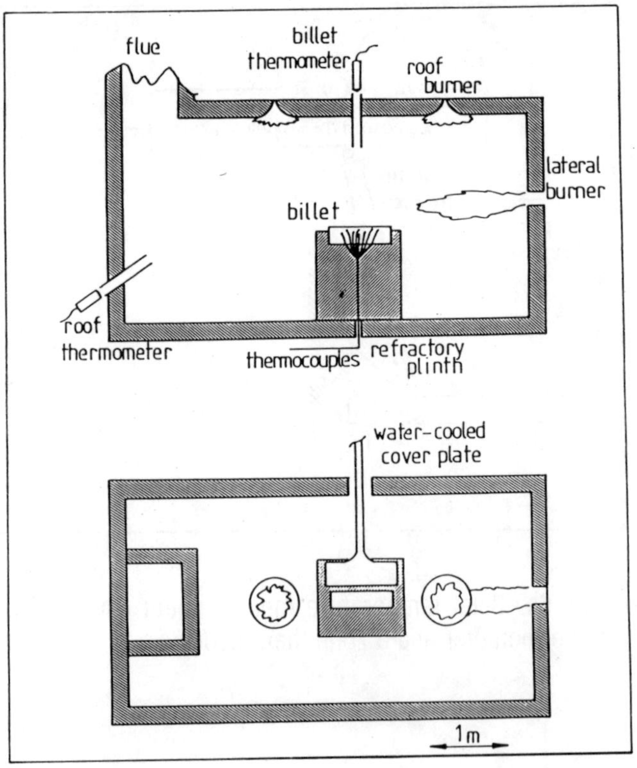

Figure 9. Furnace layout for trial at CSM.

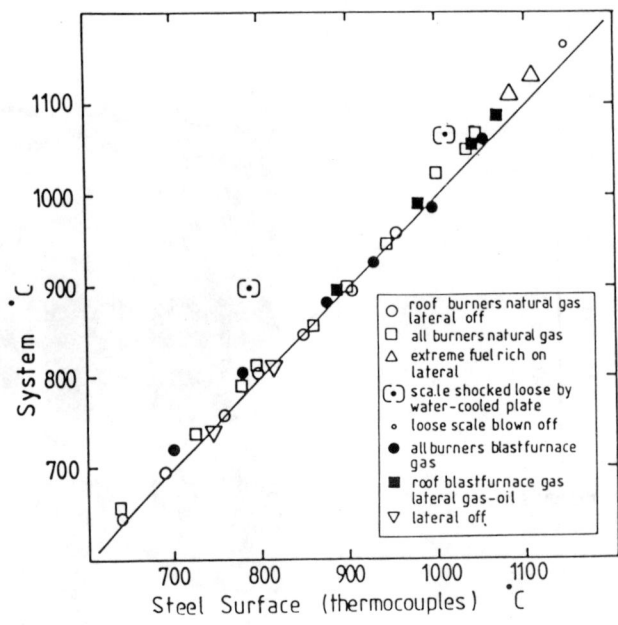

Figure 10. Composite sample data from CSM trial.

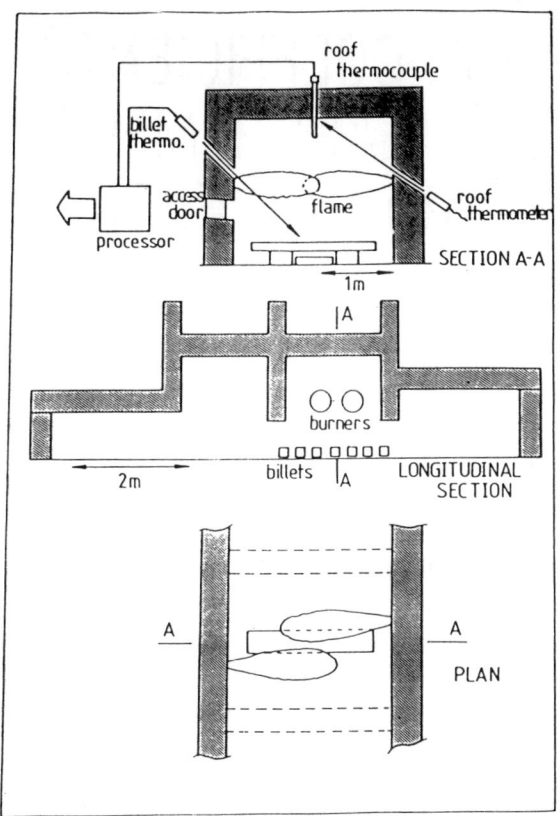

Figure 11. Furnace layout for trial at MEFOS.

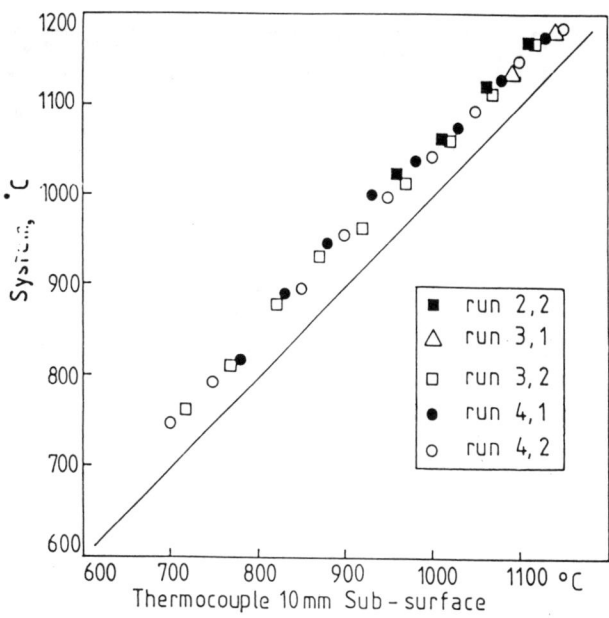

Figure 13. Composite sample data from MEFOS 1983 trial referenced to thermocouples 10 mm below surface of billet.

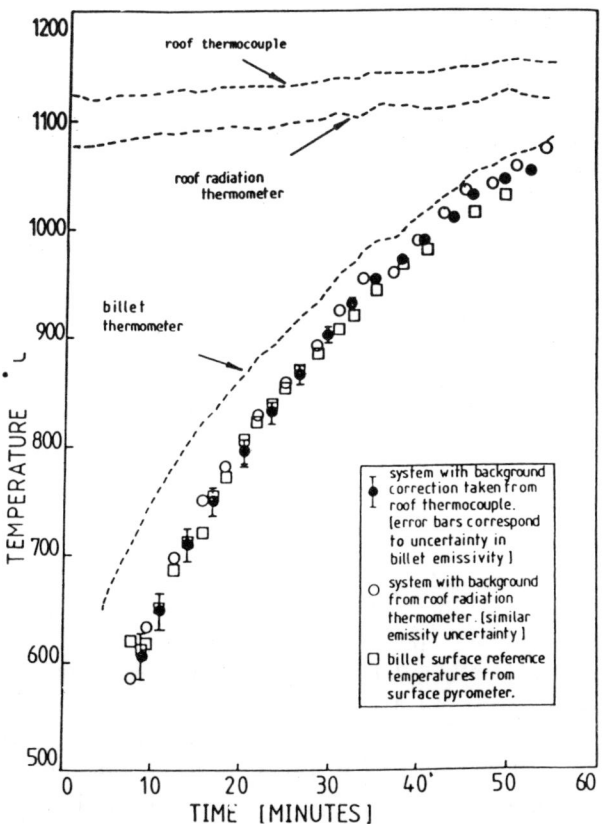

Figure 12. Specimen results from MEFOS 1982 trial.

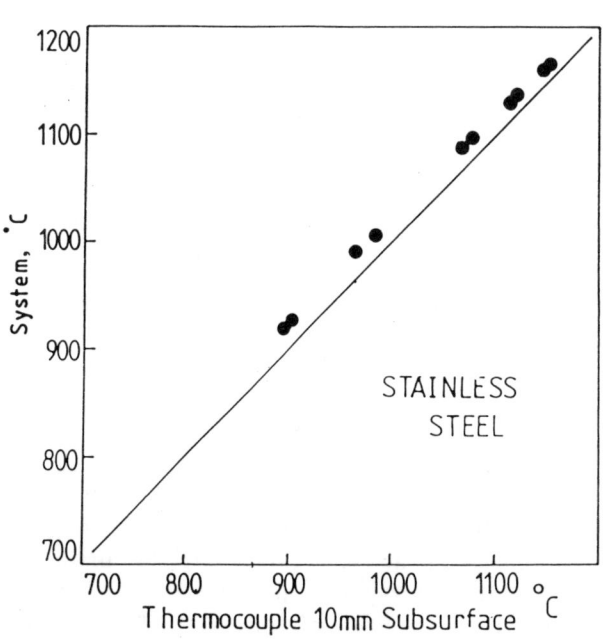

Figure 14. Run on stainless steel billet, MEFOS 1983 trial.

THE EFFECTS OF ABSORPTION COEFFICIENTS ON RADIOMETRIC MEASUREMENTS IN PROCESS FURNACES

R.P. Madding, R.P. Bruno and G.J. Burrer ■ Inframetrics, Inc., 12 Oak Park Drive, Bedford MA 01730

Particulates present in the firebox of a process furnace influence radiometric measurements made for the purpose of inferring furnace tube temperatures. It has been hypothesized that they influence the radiometric data as an exponential function dependent upon the pathlength in the firebox. Two sets of *in situ* radiometric data are reviewed. The limited data available made a reasonable first-order fit to the hypothesized mathematical relationship.

INTRODUCTION

Infrared imaging radiometers operating in the 3 to 5 and/or 8 to 12 μm spectral bands have been used in refineries for maintenance diagnostic applications for many years.[1,2] They are used primarily to locate hot spots on furnace exterior surfaces which can be indicative of the initial stages of a refractory breakdown and to find pending problems in electrical distribution systems. They are also used routinely to identify sites throughout the refinery where energy losses might be excessive. Even though some quantitative techniques are used to indicate the severity of the problems found in the electrical distribution system, for the most part all of these applications involve a minimal amount of quantitative radiometric analysis. The problems are normally found and adequately diagnosed through qualitative heat patterns.

These far-infrared imaging radiometers are also useful as diagnostic tools to observe, though inspection ports, the status of the process furnace firebox. A qualitative assessment of pending need of maintenance on the interior refractory liner, the burners, and the tubes can be made. In regard to the tubes, quantitative radiometric information is also of interest, if reasonable accuracies in tube temperatures can be derived through computations based upon the far infrared *in situ* radiometric measurements.

When there are uncertainties as to the true value of surface emittance, the fractional change in spectral radiance at temperatures which exist within a process furnace firebox, suggests better accuracies can be obtained using radiometers operating at the shorter near-infrared wavelengths. However, DeWitt[3] notes that when the surface temperatures are being measured radiometrically in the presence of background temperatures, which are notably hotter than the surface of interest, the advantage of using the shorter wavelength instruments for accuracy improvements no longer exits. In fact it appears there is some accuracy advantage to using the far-infrared portion of the spectrum.

The *in situ* temperatures of the tubes in a process furnace are of interest to those who are responsible for the furnace operation. Reasons are:

1) Localized overheating is the predominant cause of tube failure. Coke tends to deposit on internal surfaces of the tubes in overheated regions of ethylene furnaces; this is caused by decomposition of the hydrocarbon fluid flowing in the tubes. A knowledge of temperature patterns along the tubes is of value for timely scheduling of de-coking operations. Imaging radiometers have also been used to monitor the tubes during

the de-coking process. With the hydrocarbon feed stopped but with steam flowing through the tubes, the imaging radiometer can detect local areas where the coke deposits are still present.

2) While the furnace is being run at peak capacity, the tubes can exceed their design limit temperatures. At these temperatures, coke can build up rapidly with tube ruptures more likely to occur in the overheated zones. This is particularly important for ethylene pyrolysis processes. In those furnaces, the tubes often operate continuously at temperatures close to the critical point where the design strength of the tube alloy falls rapidly with small temperature increases.

3) Some process conditions result in two-phase flow within the tubes. The upper portion of a horizontal tube supporting two-phase flow is often hotter where the rate of heat transfer from the hotter tube to the fluid through a gaseous phase is less than the rate of heat transfer through a liquid phase. Under these conditions the temperature patterns around the tubes could indicate the extent to which two-phase flow is present.

This paper reports the findings of two initial studies made to determine the radiometric significance of the particulates within the firebox of a process furnace. The studies suggest that the particulates must be considered when quantitative radiometric measurements are to be made through any appreciable firebox path length (say > 6m (20 ft)) for the purpose of deriving tube temperatures. The two studies were made on the same process furnace using a narrow spectral band filter ($\approx 0.2~\mu$m wide) centered at 10.8 μm. The furnace was being fired with No. 6 fuel oil containing "catalyst fines" in the first study and with 95%-98% LPG and butane in the latter. For convenience to the reader, this paper repeats much of the tutorial content of another paper describing the first study[2].

CONFIGURATION OF A TYPICAL PROCESS FURNACE

The type of process furnace of interest is shown schematically in Figure 1. Crude oil rapidly flows (often 11m^3/min (3000 gpm)) to the inlet of tubes located along the walls of a firebox

which often operates at a temperature in excess of 1000°C. The tubes are serially connected so the vaporized crude oil gas makes multiple passes prior to leaving the furnace. While in the furnace tubes, the gas temperature increases by some 75°C. Figure 2 shows the horizontal tubes along the furnace wall and the burner flame on the end wall. Figure 3 is a photograph taken through an inspection port of an operating process furnace.

Coke deposits can build up on the inside of the tubes. As previously stated, this is most likely to occur when the tubes reach excessively high temperatures. Once the deposits begin to form they restrict the heat transfer from the tube to the oil, because of the relatively low thermal conductivity of carbon. This results in additional heating of the tube, and eventually localized zones along a tube can reach a temperature which might result in tube rupture. Figure 4 shows how the coke becomes deposited on the side of the tube facing the firebox. Figure 5 is a photograph of a tube section with excessive coke deposits.

The furnaces are designed to run continuously for several years prior to scheduled shutdown for major maintenance. Therefore diagnostic techniques, which are well suited to *in situ* appraisal of the performance status, are significant to those responsible for the operation of the furnaces.

TRANSIENT REFERENCE TARGET MEASUREMENT TECHNIQUE

A method commonly used to determine process furnace tube temperatures utilizes a special reference target that is inserted into the furnace through an inspection port. This target is usually fabricated with the same material used for the furnace tubes so after adequate carbon build-up and oxidation its emittance can be assumed to approximate the emittance, ϵ_t, of the furnace tubes.

In the transient method, most commonly used, the radiometer looks into the furnace through an inspection port at a tube location of interest. The system *isotherm* is set to mark the radiance level for this tube location. The reference target is then inserted into the furnace and observed with the preset imaging radiometer. When the reference target also triggers the *isotherm*, a temperature reading of a thermocouple, which is imbedded in the surface of the target, is quickly taken. The radiometric analysis is reduced to the simple expression:

$$T_t = T_{tc} \qquad (1)$$

that is, the tube and thermocouple temperature are equal. System calibration errors are eliminated along with the emittance parameter uncertainties. The reference target will typically reach its steady temperature in a few minutes after insertion into the firebox. Therefore an error may result from thermal lag of the thermocouple[5].

The basic elements involved in the radiometric chain are illustrated in Figure 6. For this transient technique to work satisfactorily, the radiometer system must be operated with a narrow-band spectral filter that blocks the spectral-line radiance from the combustion gasses. With the filter, the radiance from the gas medium (atmosphere) is normally neglected. Note, however, the studies reported in this paper suggest particulates within the firebox may cause a meaningfull measurement error if the line-of-sight distance from the reference target and the tubes is significantly different (say > 6m (20 ft)).

Some form of ambient chopping as illustrated in Figure 6 is a necessary part of the radiometer *dc-restore* function which serves to keep the radiance reading constant in the presence of a changing average radiance level within the radiometer field-of-view. Figure 7 depicts the transient time function showing the crossover where the reference target temperature equals the tube temperature.

STEADY-STATE TARGET MEASUREMENT TECHNIQUE

The steady-state method is similar to the transient method. The same type of target is used with a thermocouple imbedded in its surface. The reference target reaches its steady-state temperature, T_r, which will of course be greater than the tube temperature. The radiance difference (*isotherm* difference) between the radiance from the tubes, $\epsilon_t N_t$, and the steady-state reference target, $\epsilon_r N_r$, is then read with the radiometer. To determine the tube temperature, T_t, both the tube and reference emittances $\epsilon_t = \epsilon_r$, as well as the system calibration function, F_{sc}, must be known.

$$T_t = F_{sc}\left[\epsilon_t\left(N_r - N_t\right)\right] \qquad (2)$$

where F_{sc} ($^{\circ}C$/*isotherm* unit) is an empirical system temperature calibration function obtained with the narrow-band filter in place using blackbody radiators; in this case with N_t and N_r representing blackbody radiance values.

Figure 8 shows the steady-state condition reached by the reference target after passing through the transient crossover. Tube temperatures derived with this technique, as compared to the transient method, are subject to the additional uncertainties of emittance values and system calibration errors. The approach is, of course, also more susceptible to errors from particulates in the line-of-sight.

HYPOTHESIZED RADIOMETRIC MODEL FOR A PROCESS FURNACE

When the firebox of a process furnace is studied with an imaging radiometer, a flowing type fog can usually be observed. The fog is apparent, though less dense, even when a narrow-band flame suppressing filter is employed. The fog is comprised in part of carbon particulates from the combustion process. These particles act as near blackbody radiators. Therefore, narrow-band spectral filtering cannot be used to eliminate their radiometric presence. It can be assumed that the particulates:

1) Attenuate the tube radiance as it propagates through the gas medium resulting in a transmission factor, τ_p;

2) Represent another independent radiance term, $\epsilon_p N_b$ which the radiometer senses, where ϵ_p is the effective emittance for a specific path length and N_b is the blackbody radiance; and

3) Scatter or reflect, ρ_p, some of the furnace background radiance, N_b, along the radiometer optical path.

The radiance terms pertaining to the energy received by the radiometer are portrayed in Figure 9. Classically a gas or particulate medium of this type can be characterized to a first-order approximation by an exponential relation dependent upon the pathlength through the medium. Simply, the radiance from the furnace, N_f, at the radiometer optical port can be expressed as

$$N_f = \tau_p \left[\epsilon_t N_t + \left(1 - \epsilon_t \right) N_b \right] + \left(\epsilon_p + \rho_p \right) N_b \quad (3)$$

where the form of τ_p, ϵ_p, and ρ_p is assumed to be

$$\tau_p = \exp \left(-a_1 \ell \right) \quad (4)$$

$$\epsilon_p + \rho_p = \left[1 - \exp \left(-a_2 \ell \right) \right] \quad (5)$$

The particulate emittance is ϵ_p, transmittance is τ_p, and reflectance is ρ_p. a_1 and a_2 are absorption coefficients and ℓ is the pathlength through the furnace firebox. Since conservation of energy requires $\tau_p + \epsilon_p + \rho_p = 1$, first-order theoretical considerations suggest a_1 should equal a_2.

The radiance terms containing the scanner self-radiance factor, N_s, shown in the lower right portion of Figure 9 have been neglected. They are compensated by the radiometer electronic functions and system caliibratioon.

ESTIMATE OF THE RADIOMETRIC PARAMETERS IN THE MODEL

Both a dual temperature reference target and an imaging radiometer were used in the two studies to obtain *in situ* radiometric data that were used to derive values for the parameters required to solve Equation (3). The radiometric data were obtained in an operating process furnace at the Gulf Refinery in Philadelphia, which is similar in type to the one previously described. The radiometer, an Inframetrics Model 525, employed an ambient temperature chopper method to obtain absolute radiometric data.

The complete dual-target radiometric system is illustrated schematically in Figure 10 where the system components are the same as those of Figure 6. Figure 11 is a photograph of the dual target which was developed by Gulf R&D Division. It shows both the imbedded thermocouples and the inlet which provides air to cool the right target. With adequate air flow the cooled-reference target reaches a steady-state

temperature lower than the tube temperatures. Consequently, the references bracket the unknown tube temperatures. The steady-state temperatures for all three radiating surfaces are illustrated schematically in Figure 12 as they might fit with a typical absolute radiometer calibration curve.

The furnace dimensions are 18.6m x 14.3m x 7.6m (61'Lx47'Wx25'H). The radiometric measurements were taken from three inspection ports located on the 16m (47') walls which also contain the furnace burners. The dual-temperature targets were observed from two ports with an approximate furnace pathlength of 1.8m (6') and 17.6m (55'). The third port was centrally located and was used to obtain measurements of the furnace background radiance, N_b.

When the first set of radiometric measurements were taken, the furnace was operating at a relatively low capacity of some 110,000 BPD (barrels per day) of feedstock. The furnace was fired with No. 6 fuel oil containing catalyst fines. To both the eye and the imaging radiometer, the combustion process appeared "clean" relative to previous observations on the same furnace with oil firing.

At the time the second set of data were taken, the furnace was operating at a higher capacity of some 125,000 BPD feedstock. The furnace was burning 95%-98% LPG plus butane.

The following measurement sequence was used to establish values for the parameters needed to solve Equation (3).

Target and Reference Emittance, ϵ_t

Absolute radiance readings were obtained for the dual-temperature reference targets in the furnace at a pathlength of some 2m (6 ft). The emittance was derived using a radiometric computer programmed with a look-up table representative of the system calibration curve. A typical curve is illustrated in Figure 12. With the program, an emittance value was selected which resulted in a computed radiometric temperature difference equal to the measured thermocouple temperature difference. Since both targets were flat and in the same orientation, the reflected background terms canceled. Using this method the emittance value of the tube material in the first study was found to be 0.89. (The reference targets were re-surfaced prior to beginning the second study. Target emittance measurements were made at the beginning of the second study; apparently before the re-surfaced targets had adequately re-oxidized and carbonised. This second set of data showed an unreasonably low emittance of ~0.15. Unfortunately, the radiometric computations were not completed while the radiometer was on site and there was no opportunity to remeasure the emittance after more carbon had deposited. The second set of emittance values has therefore been disregarded as irrelevant for the purposes of these initial studies.)

Gas Transmittance, τ_p

The gas medium transmission factor was determined from a measurement of the radiance differences between the hot and cold reference targets at two different furnace pathlengths; 1.8m and 16.8m (6' and 55'). A telescope was used with the radiometer to assure that the dual reference target areas appeared as extended surfaces at the 16.8m (55') distance. The furnace gas transmission factor was assumed to be 1 at the 1.0m (6') distance. In the first study, at 16.8m (55') the observed radiance difference was reduced by a factor near 0.6 from the radiance difference observed at 1.0m (6'). In the second study the factor was found to be near 0.5. Using Equation (4), an average value of the absorption coefficient, a_1, for both studies is approximately 0.033m^{-1} (0.01 ft^{-1}).

Background Radiance, N_b

The same radiometric computer with a calibration curve look-up table was used to convert the absolute background radiance readings to a representative background furnace temperature. The radiance readings were taken through an inspection port centrally located in the furnace wall. From this position the path length through the furnace to the far wall surface was in the order of 18.3m (60'). The temperatures were observed to be quite uniform. Figure 13 and 14 show the difference in appearance of the flame patterns for the two fuels as photographed from the video tape recorded with the 10.8 μm filtered imaging radiometer. It is interesting to note that very little flame pattern exists when view at 10.8 μm. As might be expected the video tape shows noticeably more flame pattern when the burners are fired with No. 6 fuel oil as opposed to LPG.

At midwall, the background temperature was found to be near 1100°C in the first study when oil fired and near 900°C in the second study when LPG fired. Figure 15 is a radiant temperature profile plotted across two burner nozzles. The plot was made with an IVS 190 Digital Image Processor from the video tape made during the second study. The left burner is on (~40<X>~60) and the right burner (~170<X>~190) is off. The flame temperature as measured at 10.8 μm is lower than the wall temperature between the burners. Figure 16 is a 30-second time plot of a point on the furnace wall just left of the active burner nozzle. It was also made with the IVS 190 from the video tape. The standard deviation obtained from a small population of 5 single-point radiance measurements at different locations on the burner wall was ~60°C.

Tube Temperature, τ_t

An absolute measurement of the tube radiance near the inspection port was made, which, without consideration of the particulates, resulted in a computed tube temperature of approximately 680°C in both studies. The radiance from the top of the tube was observed to be greater than the radiance from the bottom of the tube. Computations indicate that the amount of change observed can be attributed to background radiance differences where the upper portion of the tube reflects the flame temperature over a relatively short pathlength and the lower portion of the tube reflects the radiance from the cooler furnace floor tubes, again over a relatively short pathlength.

Radiance from the Furnace, N_f

Specific data were collected on video tape in the second study to assess the basic need for an algorithm to account for the pariculates along the line-of-sight in the firebox. The 10.8 μm filtered radiometer was set up to simultaneously view nearly the entire length of horizontal tubes on one wall of the process furnace. It was configured to record absolute radiance values. Figure 17 is a photograph of the scene obtained on playback from the video tape. The reference targets can be seen in the upper left portion of the picture. The scene provides radiance data from the tubes over a continuous line of sight of some 4.3 to 18.3m (14 to 60 feet) through the furnace firebox. The tubes themselves can be assumed, with confidence, to be at a nearly constant temperature.

Figures 18, 19, and 20 are temperature/time plots of apparent tube temperatures from this scene over a 50 second-time period at three distances of approximately 4.3m, 9.1m and 18.3m (14′, 30′ and 60′), respectively. Figure 21 is an image processed version of Figure 17 showing a brightened line along the second tube from bottom of the photograph. The temperature profile from the radiance observed along the brightened line is plotted in Figure 22. Error bars have been superimposed on Figure 22 to show the maximum, minimum, and average temperature change with time, using the data plotted in Figures 18, 19, and 20. All of these plots were made with instantaneous data point sampling through the IVS 190. The radiometric algorithm in the IVS was calibrated using an assumed tube temperature of $680^{\circ}C$ at the 4.3m (14′) distance (neglecting path length effects) and a far wall furnace temperature of $900^{\circ}C$ (other *in situ* data suggests a wall temperature of $925^{\circ}C$).

The hypothesized exponential algorithm was not entered into the IVS so the apparent tube temperatures plotted do not account for any of the effects of the particulates in the firebox. Further, without a correction algorithm for the radiance contributed by the particulates at the two IVS calibration set points, it should be expected that the temperatures plotted in Figure 22 are offset lower than the actual temperatures representative of the total radiance, N_f, being measured. The plot shows a continual increase in the apparent tube temperature as the firebox pathlength increases; from $677^{\circ}C$ in the near field to $750^{\circ}C$ in the far field. As stated previously, the tubes can be assumed to be at a constant temperature. Therefore, the apparent increase suggested by the plot is unreasonable. The data unquestionably suggest an algorithm is needed to account for the particulates.

With the "approximated" parameter values derived from the two initial studies, as given in the four previous subsections, the validity of the hypothesized exponential function was assessed at three pathlength distances; near at 4.3m (14′), mid at 9.1m (30′), and far at 18.3m (60′). When Equation (3) is solved, the range of N_f measured values is reasonably close to the computed values. This is particularly true when recognition is given to the offset in the measured values which results

from the selection of uncorrected calibration temperatures. The computed temperatures for the radiance, N_f, when corrected for the particulates are also shown in Figure 22. It appears Equation (3) represents a reasonable expression of the total emitted and reflected radiance, N_f, reaching the radiometer from both the tubes and the particulate (and/or combustion gases) sources. However, the spread in measured values is large due to the continually changing conditions within the firebox as illustrated by Figures 18, 19, and 20. These fluctuations alone represent an uncertainty in tube temperature of at least $60^{\circ}C$.

RECOMMENDATIONS

The radiometrics within the firebox are quite complex. It is therefore necessary to obtain more empirical data in order to truly assess the validity of the simple model hypothesized. One future study should evaluate a set of narrow-band spectral filters centered at different wavelengths across the 3 to 5 and 8 to 12 μm spectral windows. The results will better assure that it is only the graybody particulates that are influencing the measurements, not the combustion gases themselves. Future work should also be concentrated on the collection of a sufficient amount of data to allow meaningful statistical analysis for quantification of the variances that occur over time and along different pathlengths. The magnitude of these variances with statistical averages in all probability will determine the ultimate value of this type of quantitative radiometric diagnostic.

CONCLUSIONS

These initial empirical findings using the hypothesized radiometric model were found to be reasonable. Additional *in situ* data are required to further test the model and make some assessment of the dependence of the radiometric parameters on furnace firing conditions. If reasonably small parameter variances in averaged values are observed over time with constant firing conditions, the technique could evolve into a viable method for routine monitoring of furnace tube temperatures.

The radiance measurements made in these two initial studies suggest that firebox particulates over a pathlength of some 15m (50′) may attenuate contrast readings by a factor near 0.6 and contribute an additional radiance term of a magnitude near 0.5 of the furnace background radiance level.

ACKNOWLEDGEMENT

The authors are indebted to J.B. McCandless, and A.G. Imgram of Gulf Research and Development Co., Pittsburgh, PA; D.C. Damin now with E.I. duPont de Nemours, Beaumont, TX; and D.P. DeWitt of Purdue University, W. Lafayette, IN for their suggestions and assistance in collecting and interpreting the radiometric data. Furnace access, the dual temperature target, and the imaging radiometer were provided by Gulf Research and Development Company.

REFERENCES

1. Imgram, A.G., "Infrared Thermal Imaging of Refinery Equipment," Proc. Thermosense V, SPIE, Vol. 371, pp. 47-54. Oct. 1982.

2. Bruno, R.P. and Burrer, G.J., "Appraising Process Furnace Tubes with Imaging Radiometers," Proc. Thermosense VI, SPIE, Vol. 446, pp. 130-136, Oct. 1983.

3. DeWitt, D.P., "Inferring Temperature from Optical Radiation Measurements," Proc. Thermosense VI. SPIE, Vol. 446, pp. 226-233, Oct. 1983.

4. Bruno, R.P., "Analytical Interpretation of Thermograms Including Digital Processing Techniques," Proc. Thermosense V, SPIE, Vol. 371, pp. 230-239, Oct. 1982.

5. Madding, R.P., "Infrared Sensors and Process Control," Proc. Thermosense VI, SPIE, Vol. 446, pp. 9-17, Oct. 1983.

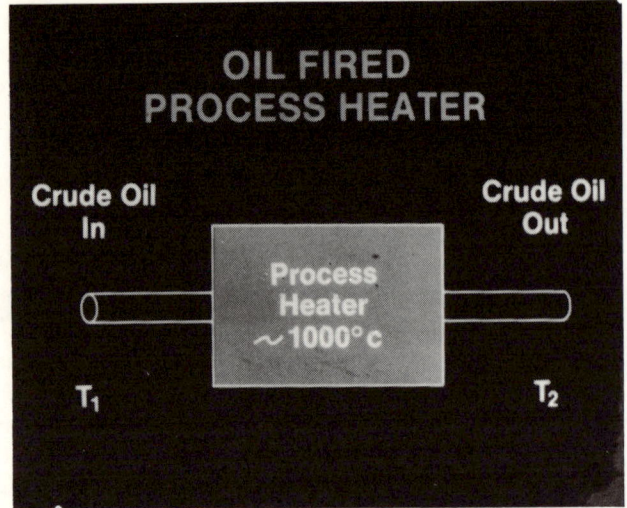

Figure 1. Schematic representation of the oil-fired process heater.

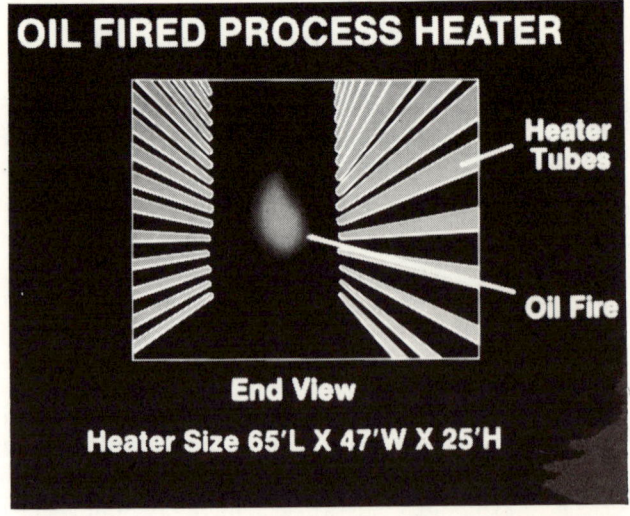

Figure 2. End view of the oil-fired process heater.

Figure 3. End view of the oil-fired process heater (photograph).

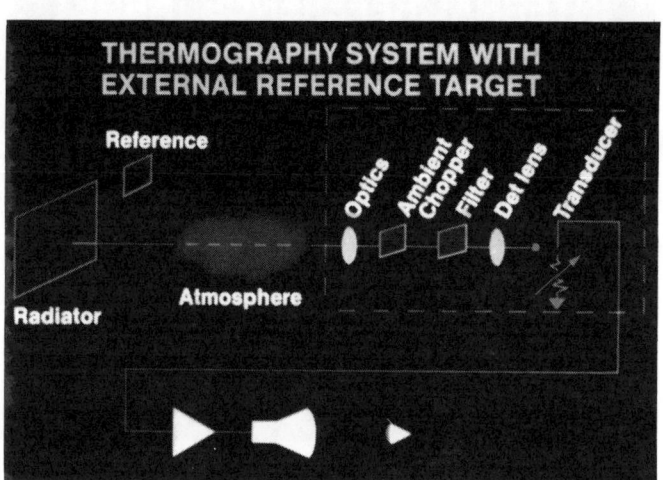

Figure 6. The thermography system.

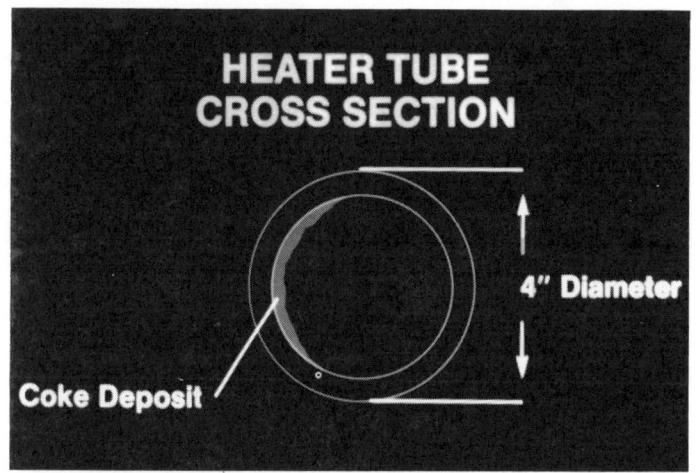

Figure 4. Coke deposit on a heater tube.

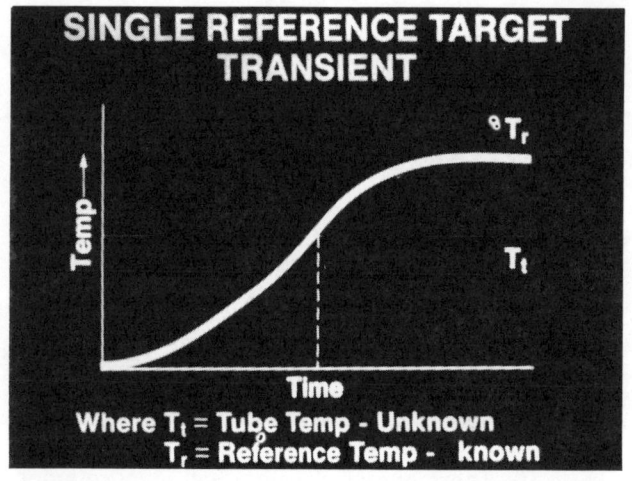

Figure 7. Transient temperature response of the reference target.

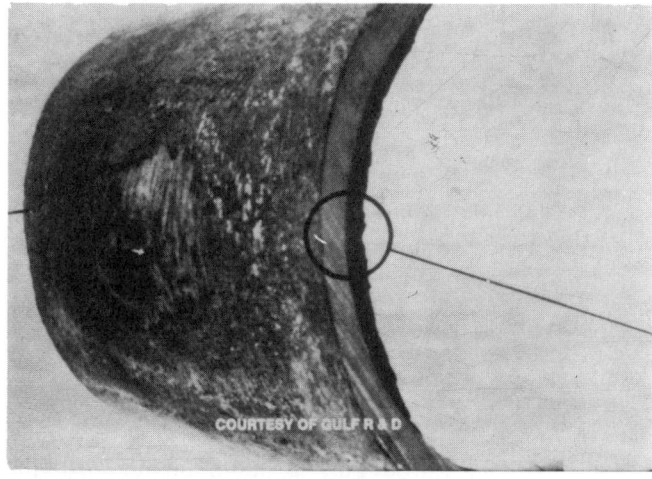

Figure 5. Coke deposit on a heater tube (photograph).

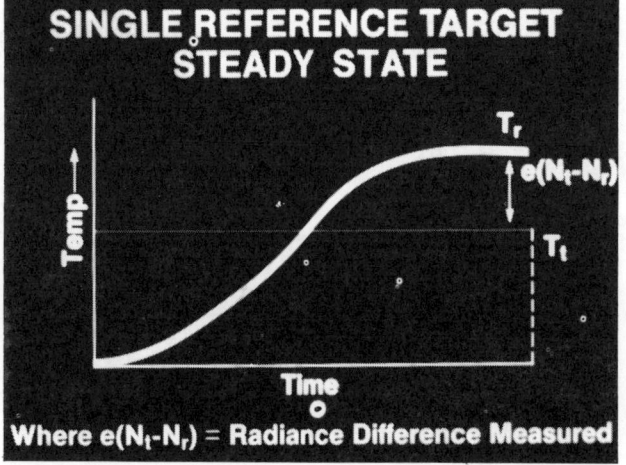

Figure 8. Steady-state temperature of the reference target.

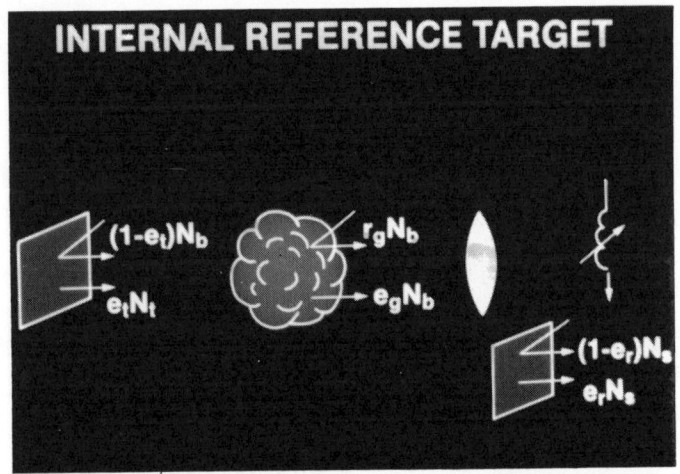

Figure 9. Internal reference target of the thermography system.

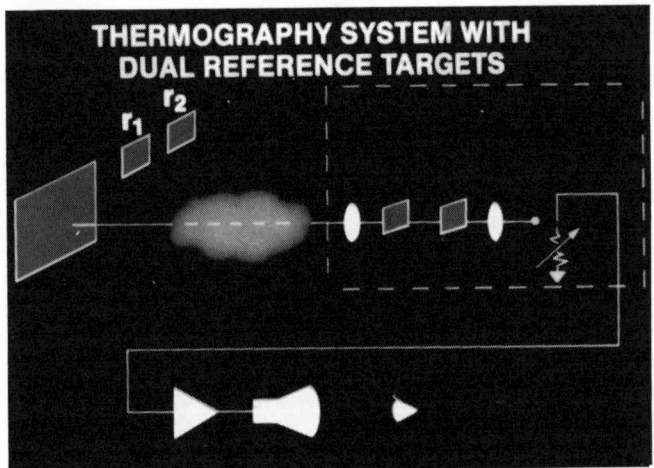

Figure 10. Thermography system with dual reference targets.

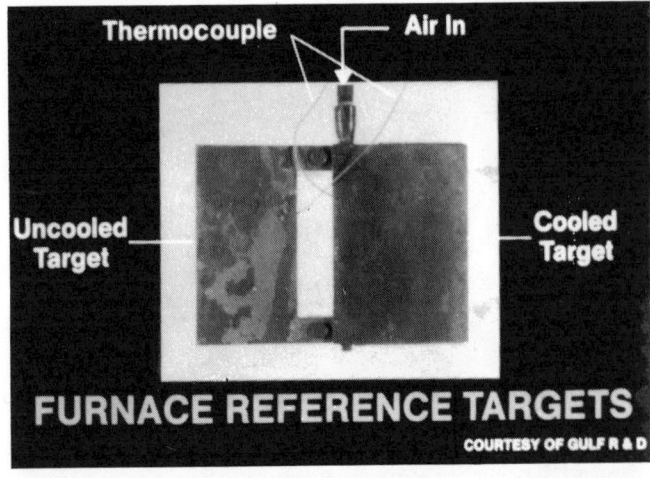

Figure 11. Furnace reference targets.

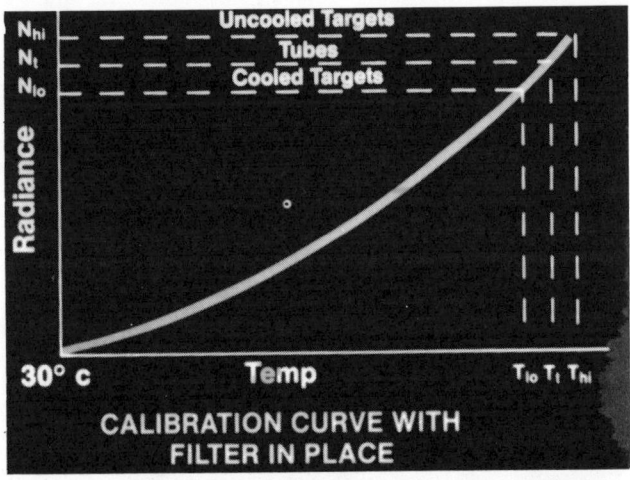

Figure 12. Calibration curve with filter in place.

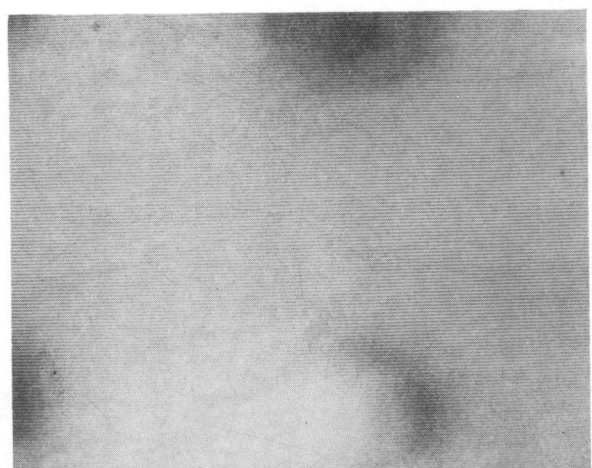

Figure 13. Oil-fired burner flame at 10.8 μm (bottom center on).

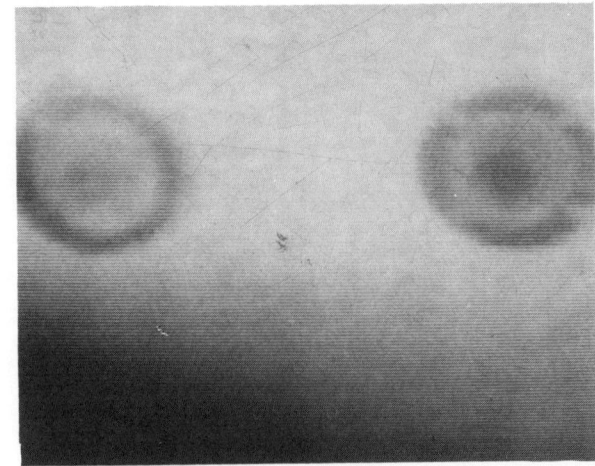

Figure 14. LPG-fired burner flame at 10.8 μm (left burner on).

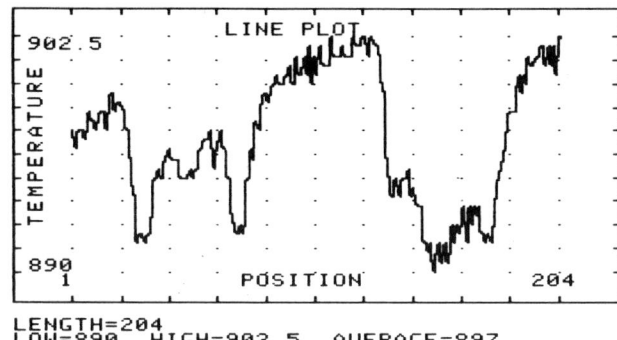

Figure 15. Temperature profile (°C) across an LPG fired nozzle at 10.8 μm (burner on at location 40 < X > 60; burner off at location 170 < X > 190).

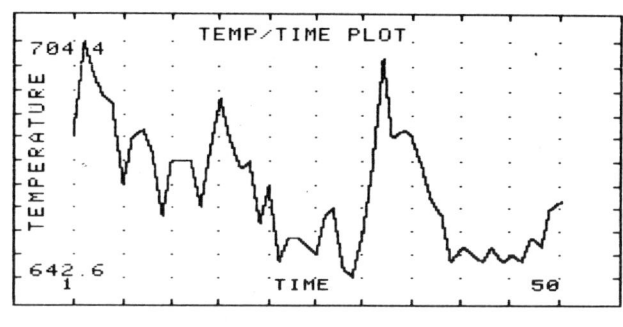

Figure 18. Temperature (°C) vs. time (s) plot of apparent tube temperature (no correction for the presence of fire box particulates) at distance of 4.3 m (14 ft) with 10.8 μm filter.

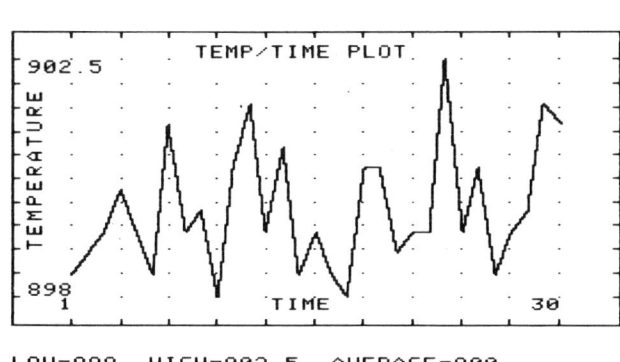

Figure 16. Temperature (°C) vs. time (s) plot near a LPG-fired nozzle.

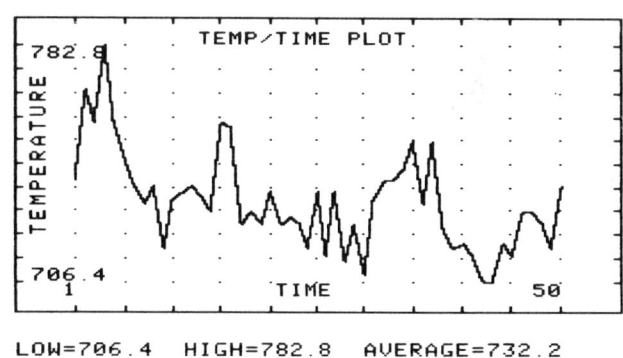

Figure 19. Temperature (°C) vs. time (s) plot of apparent tube temperature at a distance of 9.1 m (30 ft) with 10.8 μm filter.

Figure 17. Thermogram of tubes as recorded with 10.8 μm imaging radiometer.

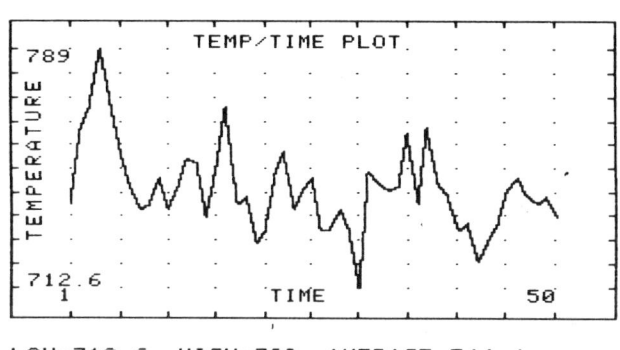

Figure 20. Temperature (°C) vs. time (s) plot of apparent tube temperature at a distance of 18.2 (60 ft) with 10.8 μm filter.

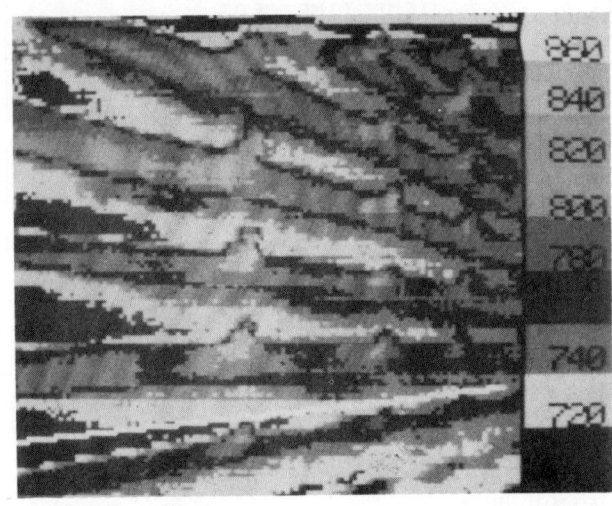

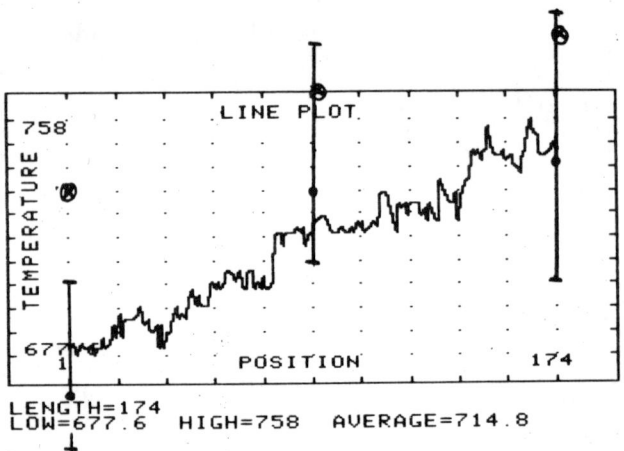

Figure 21. Processed image of Figure 17 with brightened line along second tube from bottom.

Figure 22. Apparent tube temperature along brightened line illustrated in Figure 21. Three vertical lines represent Max, Min and Ave data from Figures 18, 19 and 20. The symbol ⊗ shows computed value.

PRACTICAL CONSIDERATIONS IN THE INFRARED MEASUREMENT OF HOT GAS, FLAME, AND TUBE WALL TEMPERATURES

E.M. Emery ■ E² Technology Corporation, Ventura, California

Practical methods for measuring the temperature of hot gases, flames, and tube wall temperatures are discussed. Proper selection of the spectral passband of the infrared thermometer is required in order to observe gas and flame temperatures. The primary errors in measuring tube wall temperatures are due to uncertainties in the surface emissivity and reflected irradiation from hotter zones in the furnace. To minimize these effects, a tube-surface blackbody target is proposed.

INTRODUCTION

The need to measure gas or flame temperature to conserve energy, improve production, or reduce costs arises in a variety of processes such as boiler control, furnace control, chemical reactors, flame hardening and welding, coal gasification, flares, jet engine discharge, and rocket plumes.

Reactants and products of combustion vary widely but for most industrial processes, the major components in the products of combustion (flame) are CO_2, CO, C, H_2O and N_2 with some processes producing additionally SO_3, SO_2, NO_2 and NO. To measure the temperature of the products of combustion using infrared radiation techniques, it is desirable to eliminate the influence of unburned fuel which may be present in the combustion zone.

One of the main strengths of the infrared thermometer is its ability to operate at different spectral bands in the infrared spectrum. This capability makes the infrared thermometer unique among temperature measurement instruments since it can literally ignore the presence of one component of a process while measuring the temperature of another. Each gaseous element or compound has a characteristic spectral transmission or absorption signature.

SPECTRAL SIGNATURES — HOT GASES AND FLAMES

By properly selecting the infrared filter for the infrared thermometer to absorb or pass a specific wavelength, the instrument can selectively measure (or ignore) the temperatures of many gaseous components. Fortunately an absorption band in the infrared spectrum of most hydrocarbon fuels occurs between 3.0 and 3.5 μm with good transmission starting above 3.5 μm. (Figure 1). The absorption band for most of the major products of combustion occur between 4 and 5 μm. (Figure 2). This separation of absorption bands provides a convenient method of filtering out the effects of unburned hydrocarbons using a filter which passes energy only in the 4 to 5 micrometer spectrum. (Figure 3). The relatively simple explanation however is only the first cut at effective flame temperature measurement. One must next consider the purposes of the measurement: to determine the temperature of the "plume" of the flame for mixture ratio or stoichiometric analysis, and to determine the total energy in a furnace or boiler available for heat transfer.

FLAME TEMPERATURE MEASUREMENT

In the case of flame temperature measurement, a spectral range should be selected which will preclude the cooler products of combustion and unburned fuel from absorbing the energy radiated by the flame. Recognizing that the absorption bands of most gases tend to broaden and shift to higher wavelengths with increasing temperature, it is possible to minimize the effects of lower temperature (cooled) products of combustion such as CO_2, CO and H_2O by selecting an infrared filter which will pass the radiant energy only from the hotter gases.

A passband of 4.5 to $4.6 \mu m$ accomplishes this purpose for CO_2 measurement. CO_2 has an atmospheric average absorption centered at $4.26 \mu m$ while at temperatures above $800°C$ ($1472°F$), the dominant absorption band averages $4.5 \mu m$. Carbon monoxide (CO) demonstrates a lesser effect at elevated temperatures. An infrared thermometer, operating at $4.5 \pm 0.05 \mu m$ is a very effective tool for measuring actual flame temperatures of furnace jets and rocket exhaust. (Figure 4).

Length of the sight path or thickness of the flame is a critical factor in flame temperature mesurement accuracy. As a rule of thumb, a "flame path" of a least 1 foot is necessary. If there is a quantity of particulate material in the flame at essentially the same temperture as the products of combustion, the sight path may be shorter.

TEMPERATURE MEASUREMENT FOR ESTIMATING HEAT TRANSFER

On the other hand, if the temperature mesurement is to be used to determine the temperature difference available for heat transfer between the gas and a surface, another approach is required. In a gas-fired boiler for instance, the cooler gases play an important role in obtaining an average or "bulk" temperature.

Heat transfer between the gas and the furnace tubes occurs by two modes, and can be expressed as:

$$q = A \left[h_c \left(T_g - T_{ts} \right) + h_r \left(T_w - T_{ts} \right) \right] \quad (1)$$

where q is the heat transfer rate [BTU/hr] by convection and radiation; h_c and h_r are the convective and radiative heat transfer coefficients, respectively, [Btu/hr—ft^2—°F]; A is the heat transfer area [ft^2]; T_g is the average temperature of the gas surrounding the tubes [°F]; T_{ts} is the average temperature of the surface of the tubes [°F]; and T_w is the average wall (refractory) or burner plume temperature [°F] comprising the surroundings of the tube.

Convective Heat Transfer Temperature Measurements

The solution of the rate equation above assumes a uniform gas temperature, T_g, surrounding the tubes, which in reality includes a mixture of hot and cooler gases. The infrared temperture measurement which averages the temperature of the hotter gases with that of the cooler gases in the vicinity of the heat exchanger tubes is the practical measurement for heat transfer analysis. The spectral range of 4.2 to $4.6 \mu m$ will present a more useful *bulk* temperature value of the gas environment around the tubes.

Care should be taken to sight the instrument on the most characterisitic location. Typically the instrument is sighted perpendicular to the burners, that is, across the path of the flames (Figure 5). Long shots should be avoided through cooler gases which are not near the tubes, such as sighting down the tube of a feed line into the furnace. An infrared thermometer operating in the 4.2 and $4.6 \mu m$ range is used for measuring bulk gas temperature in boilers.

Radiant Heat Transfer Temperature Measurement

To obtain the value of T_w one must decide if the burner or the refractory temperature is the dominant driving force for the radiant heat transfer. In large boilers where the burners are separated from the length of the tubes, it may be necessary to use the flame temperature as a radiant source at one end of the furnace and the temperature of the refractory at the other end. Flame temperature is measured at $4.5 \pm 0.05 \mu m$ while the refractory temperature requires a different wavelength as described in the following section.

TUBE-SURFACE TEMPERATURE MEASUREMENT

Temperature measurement of the tube surface for use in the heat transfer rate equation presents a problem with the opposite requirements. One now must receive energy from the tube surface without being influenced by the surrounding gases or energy reflections from hotter sources such as flames or refractory.

Temperature measurement of the tube surface (T_{ts}) is accomplished by using an infrared thermometer in the spectral range of 2.2 ± 0.05 micrometers, a very narrow spike filter centered at the peak transmission point of most reactants and products of combustion, or at 3.9μm if no sulfur or nitrogen oxides are present in the products of combustion. Also 3.9μm shows a slightly higher absorption for unburned methane and propane.

The search for this spectral wavelength required a study of the spectral signature of each reactant and compound resulting from combustion and selecting a common spectral range from Table 1 which indicates infrared transmission at 2.2 and 3.9μm through various gaseous components which may be present in the boiler or furnace. The narrow spectral range, however, solves only half of the problem. All of the radiant energy (E_{tot}) radiating from any source incident on a surface may be reflected (E_{ref}), absorbed (E_{abs}) or transmitted (E_{tr}) according to the following conservation equation:

$$E_{tot} = E_{ref} + E_{abs} + E_{tr} \tag{2}$$

Since the tubes we are viewing are solid and not transparent, the transmission term, E_{tr}, is zero. From Kirchhoff's law we know the fraction of absorbed energy, E_{abs}/E_{tot}, and the fraction of the emitted energy compared to a blackbody will be equal. It is the emission or emitted radiant energy one wants to measure in order to determine the temperature of the radiating surface. This emission term varies with material, surface condition, wavelength, angle of incidence and temperature. Only a perfect blackbody radiator eliminates these variables and presents a total emitted energy proportional to the 4th power of the absolute temperature of its surface.

It is necessary to minimize the energy reflected from the convex surface of the tube and from other sources which may be hotter than the tube surface to achieve a true temperature reading.

A direct approach to solving the reflection and variable emissivity problems is to attach a blackbody target to the tube at the time of tube

Table 1. Transmittance of Selected gases as
2.2 and 3.9 μm

		Approximate Transmittance (%)	
Gas		2.2 μm	3.9 μm
Ammonia	NH_3	87	87
Carbon Dioxide	CO_2	95	95
Carbon Monoxide	CO	96	99
Hydrogen Nitrate	HNO	95	93
Hydrogen Sulfide	H_2S	97	96
Methane	CH_6	97	94
Nitrogen	N_2	100	100
Nitrogen Dioxide	NO_2	92	80
Nitric Oxide	NO	97	97
Oxygen	O_2	100	100
Propane	$CH_3CH_2CH_3$	97	90
Sulfide Dioxide	SO_2	97	90
Water Vapor	H_2O	100	100

assembly or furnace turnaround; such an approach introduces less error than the attachment of a thermocouple on the tube surface. If the absorptivity were equal to unity, there would also be zero reflection when using the target. With zero reflection and energy emitting from a nearly perfect radiator, the infrared thermometer would directly indicate the true temperature. Equation (2) now becomes

$$E_{tot} = 0 + E_{abs} + 0 \qquad (3)$$

and hence the only energy received by the infrared thermometer is that emitted by the tube. With blackbody targets attached to key tubes, in a reformer furnace for example, the problem of gathering reliable and repeatable temperature data is reduced to good procedures and discipline.

Is such an obvious simple approach really a practical solution to the furnace tube measurement problem? (Figure 6). A small target of the same material as the tube, less than 2" in diameter, may be multiple fillet welded to the tube. The outside surface of the target has been machined with concentric narrow V-shaped grooves creating a near-blackbody radiator effect. Emissivities of 0.99 are possible with this technique. Indeed, laboratory data indicates that radiation from a 1500°C source causes less than 7°C change in infrared thermometer reading while rotating the blackbody through a 90° arc. (Simulating radiation from different sources.) Although subtle differences have been incorporated in the tube-surface blackbody target, the concept is essentially the same as the "folded conical" blackbody radiator.

This simple, practical technique is not without cost. As a thermocouple is an intrusive device, so is the tube-surface blackbody target. It adds metal thickness and changes the convective heat transfer, which will of course affect the temperature of the target. On the other hand, the target enhances and stabilizes the emissivity in the area of the attachment.

EFFECTS OF THE TUBE-SURFACE BLACKBODY TARGET

Compensating errors could reduce the intrusive effect on accuracy of measurement.

1) Increased metal thickness or imperfect weldment, leaving a gap between the target and tube wall, will increase the resistance to heat transfer and hence decrease heat transfer, raising the surface temperature.

2) Improved convective heat transfer (the fin effect) will improve heat transfer, but only minimally.

3) Improved absorption as compared to the tubes due to higher emissivity will raise the surface temperture. This will also be a secondary effect.

4) Because the target is not a perfect radiator it will have an emissivity of less than unity, probably 0.97 to 0.99. This will tend to lower the surface temperature reading.

The greatest risk is in not having a good thermal bond between the target and the tube. Item 4) is a compensation error which tends to offset the errors listed in Items 1), 2) and 3). Hence, we conclude that the target very likely introduces less intrusive error than a typical thermocouple installation.

For large furnaces, up to 60' wide, a tube-surface blackbody target of near 2" diameter is recommended to be able to measure the temperature in the center of the furnace. At 30' the spot size resolution can be 1.2 inches, well within the boundaries of the target. Much smaller targets can be used for shorter distances.

Welding of the target to the tube is recommended for good thermal contact and ease of fabrication. The grooves should not be cut directly into the tube wall to avoid stress concentrations. The mechanical configuration of the target will minimize the effects of aging on its emission properties. Corrosion and scale formed on the surface of the grooves will not appreciably alter its emissivity properties.

Initial installation may be in conjunction with thermocouples installed between the target and the tube. For the life of the thermocouple, correlating data may be recorded to further refine the accuracy of the measurement system.

The difference of integrated gas temperature T_g and the tube surface temperature T_{ts} provides the temperature difference which drives the convective heat transfer mechanism.

SUMMARY

To measure the temperature of flame or plumes or the very hot gas in a furnace or reactor, the spectral range of $4.5 \pm 0.05\mu$m should be used. The E²Technology Model 7000FM, Pulsar II, is one such commercial instrument suited for this application.

For practical heat transfer determinations, a spectral range of 4.2 and 4.6μm gives a more integrated picture of the bulk temperature driving potential for the heat transfer process. Our Model 7000BT is appropriate for this application.

Even in large tube furnaces, small (less than 2" diameter) tube-surface blackbody targets provide useful and repeatable temperature data when observed with an instrument of high optical resolution, filtered to ignore products of combustion. Our Model 7000FM-HR is appropriate for the application.

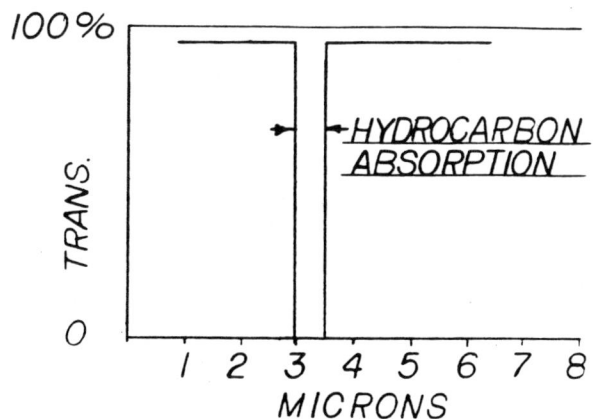

Figure 1. Idealized transmission spectra of typical (unburned) hydrocarbon fuels showing an absorption band at $3.0 - 3.5\mu$m.

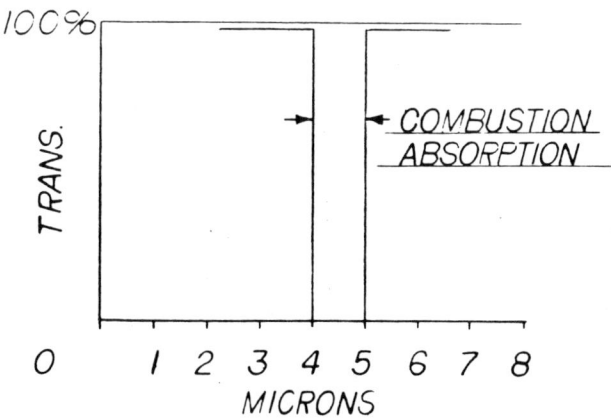

Figure 2. Idealized transmission spectra of major products of combustion showing an absorption band at $4-5\mu$m.

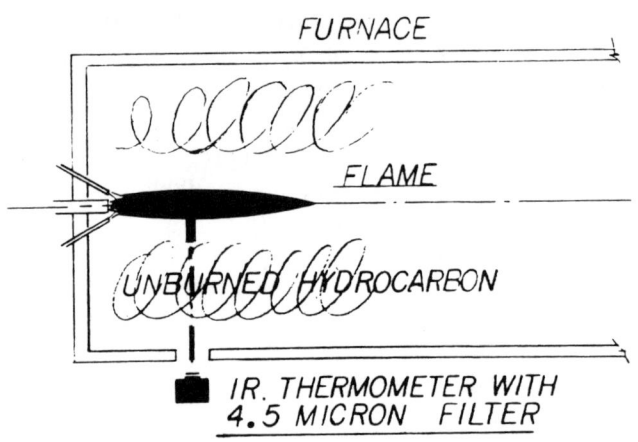

Figure 3. Using infrared thermometer filtered at 4.5μm to view a flame through unburned hydrocarbons.

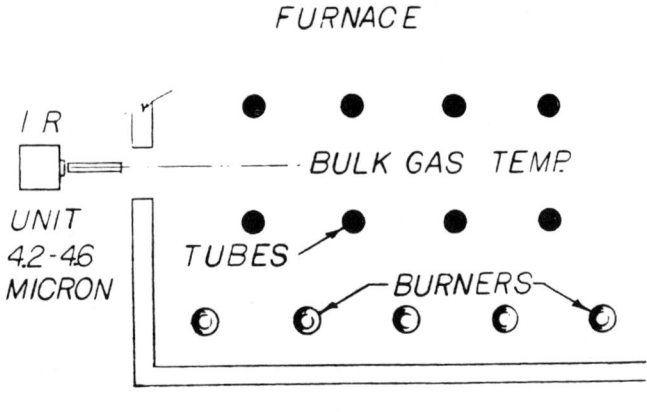

Figure 5. Sighting on tubes in a furnace.

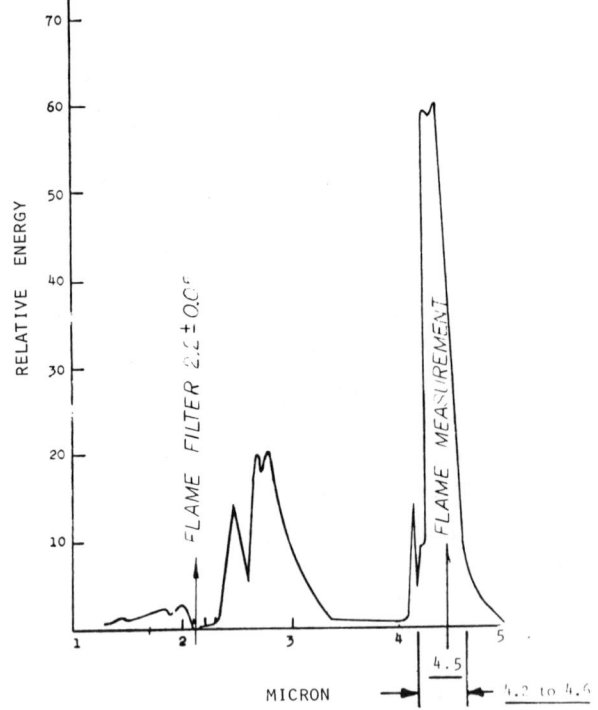

Figure 4. Emission spectra from a flame.

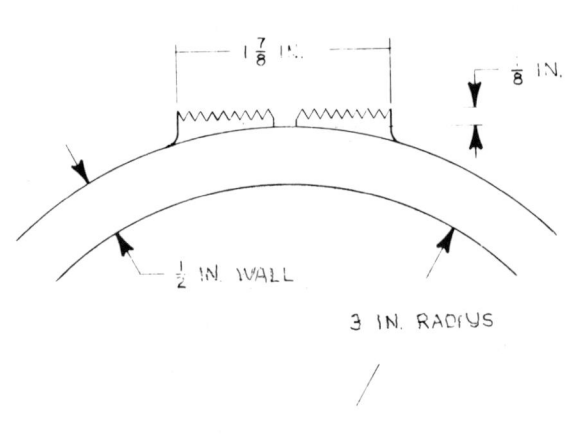

Figure 6. Schematic representation of the tube-surface blackbody target.

EXAMINATIONS OF THE THERMAL CONDITIONS OF THE CATALYTIC OXIDATION OPERATIONS

K. Welther, I. Zakarias and R. Csikos ■ Hungarian Oil and Gas Research Institute
H-8200 Veszprem, Jozsef Attila u.34, Hungary

Catalytic oxidation processes have been developed in which solid catalysts are employed. Diffusion of the reactants and products, reactions on the catalyst surfaces, and heat transfer all occur simultaneously. Concentration and temperature gradients throughout the catalyst must be determined to calculate the effectiveness of the process. Knowledge of the properties of the fluids and the physico-chemical characteristics of the catalyst are needed to perform heat and mass transfer calculations. Temperatures are measured with optical pyrometers in order to control the level of heating and the temperature levels across the catalyst surfaces.

INTRODUCTION

Both economic and technological factors dictate that certain process equipment or catalyst surfaces should be heated on only their surfaces (1-4). In such cases, only radiant heat transfer (5) is applicable. Such a conclusion is based on extensive prior literature (6-11), on numerous tests with industrial radiant heaters as observed by MA'FKI personnel over a period of many years, and on a specific research program that explored modifications of the emissions with emphasis at far infrared wavelengths.

EXPERIMENTAL PROGRAM

A specific radiant heater was developed as shown in the photograph, see Figure 1. The ceramic support had a surface of 157 × 242 mm resulting in a nominal surface area of 37994 mm². A radiant plate was formed from six identical sections whose dimensions were about 80 × 80 mm; each had a thickness of 13 mm. The six sections were compactly joined together using a strip of heat-resistant mineral fiber. Other portions of the burner, the coating, the mixing chamber, and the apron were constructed from heat-resistant stainless steel.

The ceramic unit plates that had a concave surface were each provided with 1500 holes. The diameter of each hole was 1.2 mm. The cross sectional area provided by these holes was 10,170 mm². Air used for combustion was supplied from the outside.

Three burners or appliances were built and tested for the measurement of infrared radiation (12). Each burner used ceramic plates that were of essentially identical size and shape.

1) Appliance No.1 had untreated ceramic plates.

2) Appliance No.2 had ceramic plates that were impregnated with chromium oxide.

3) Appliance No.3 employed a new surface currently being patented and based on extensive research conducted by MA'FKI.

The objectives of the tests being described were to evaluate and compare the performance of the three appliances. Temperature measurements were of course critical for such an evaluation. Tests were performed as specified by MSZ/Hungarian Standard/11423/3-73 concerning mechanical properties, environmental regulations, safety, power and operation engineering, manageability, construction and durability. The present paper is essentially restricted to reporting the results of temperature measurements and thermal conditions on the catalytic surfaces or burners.

Table 1 indicates important data for the mixtures of propane and butane used as fuels in the tests.

Table 1.

Main Characteristics of Testing Gases

Testing gas	A	B	C
Heat of combustion kcal/m^3	26340	27255	31590
Heating value kcal/m^3	24250	25165	29480
Relative density	1.790	1.762	2.090
CO_2 max. vol.%	13.8	13.9	14.2

The surface temperatures of the radiant burners were measured using an optical pyrometer of Czechoslovakian Pyronet type suitable for the temperature range of 150-2300 °C. The density of heating stream and temperature distribution of the burning were tested using infrared thermograms.

Table 2 describes the instruments used in the infrared tests of this investigation. A gas chromatographic instrument equipped with a flame-ionization detector was employed to determine the hydrocarbon content of the combustion gases from the burner. Carbon monoxide in the exhaust gases was first converted to methane in a catalytic reactor, and the methane content was then measured. A dual channel/Porapack and Molfilter/gas-chromatograph supplied with a thermal conductivity cell was employed for measurement of the CO_2, N_2, and O_2 content of the combustion gases.

Table 2.

Infrared Scanner Used

	Measuring of heating stream density	Measuring of temperature	Moisture content of air
Type	PETROSCANNER	THERM 3200	THERM 2245
Manufacture	AGA Infrared SYSTEM SWEDEN	Albrohn Mess-und Regelungs- Technik/Germany/	
Temperature range of operation	-30 °C to 70 °C	0 °C to 40 °C	-30 °C to 100 °C
Temperature range of measuring	-20 °C to 2000 °C		
Precision of measuring		±0.5%	±1%
Detector	InSb		
Range of sensitivity	2.8 - 5.4 um		

During the tests performed with radiant burners, the pressure was changed in the range of 7000-9700 Pa, loading in the range of 3.44-4.24 kW and the excess of air in the range of 1.13-1.39.

RESULTS

The results of Tests A, B, and C, as are described in Table 1, are summarized in Tables 3 and 4. Tests with the untreated ceramic No. 1 indicated that environmental, safety, manageability criteria were not met, and they hence are not reported in the tables. For devices No. 2 and No. 3 however, all desired criteria were met. For appliance devices No. 2 and No. 3, the combustion gases contained only 0.007 volume % and 0.005 volume % respectively of carbon monoxide; based on environmental standards, 0.015 volume % is permissible (with a 12.3 volume % of carbon dioxide). The non-oxidized hydrocarbon content realized with appliance No. 2 was 0.04 volume % (when obtained in an undiluted condition); for appliance No. 3, the content was even less, namely 0.007 volume %.

In tests with appliance No. 2, the surface temperature was in the range of about 840 to 870 °C. For appliance No. 3, the temperature range was 810 to 870 °C. The surface temperature changed by less than 100 °C in the time period investigated. For both appliances No. 2 and No. 3, re-ignition of the flame did not reoccur.

Figure 2 shows the thermogram of the surface of the untreated ceramic device (No. 1) during a typical test. Figure 3 is a photograph for a typical run made with appliance 3 with normal fuel loading and after several hours of operation when equilibrium conditions were realized. Thermograms are shown in Figures 4 and 5 for appliances No. 2 and No. 3. The thermograms were used to calculate the density-temperature patterns of appliances No. 1, No. 2, and No. 3. Integration techniques were employed for uniform heat fluxes resulting in Figures 6, 7, and 8 respectively. These results indicated significant differences for the three appliances:

1) Appliance No. 1 radiated over the entire range of colours, +4 to -4 which covers the range of temperatures from 892 to 608 °C; 7% of the surface, designated in Figure 6 as "a", had lower radiation.

2) Appliance No.2, as shown in Figure 7, radiated in a wide range of colours (corresponding to a wider range of temperature). 9.5% of the surface, also designated as "a", showed a lower level of radiation.

3) Appliance No. 3 radiated however in a range of temperatures from 813 to 606 °C (or range of colours of 0 to 3). Secondary radiation however occurred in the range of colours of -1 to -4; this secondary radiation corresponds to 26% of the radiatory ceramic surface and is designated in Figure 8 as "b".

CONCLUSIONS

The results of this investigation indicate that the design and operation of combustion devices are important for exhaust gas purification (i.e. meeting environmental standards) in various plants such as fermentation, sewage, or pulp mill units. The impure or waste gases from such plants can be processed satisfactorily.

The special combustion device developed by MA'FKI was found to be highly effective and better than earlier devices. This special device indicated the following:

a)   Waste gas streams can be processed to meet environmental and safety standards. This device was superior to other devices tested.

b)   Based on thermograms and other similar data, the new device produced less radiation, and during operation it was darker in appearance indicating a lower temperature. The radiation was in a narrower band.

c)   Apparently less local overheating (or fewer hot sites) were present, i.e. a more uniform temperature was occurring on the surface.

d)   Secondary radiation was caused to a considerable extent by metal fittings.

The catalytic activity of the surface is important in measuring and regulating the operation of the combustion devices.

LITERATURE CITED

1. Welther, K. and L. Szepesy, DECHEMA Monograph, Verlag Chemie (1976).

2. Hungarian Patent 171,795.

3. Welther, K. and Zs. Sebestyen, VEAB Digest III, 211 (1978).

4. Hungarian Patent 165,959.

5. Radcliff, S.W. and R.G. Hickman, Diffusive Catalytic Combustors, J. Inst.Fuel, 208 (1975).

6. Rudowski, G., " The Infra TV and Its Application, " Technical Publishing House, Budapest (1982).

7. Benkö, I., "Characteristics of Infra-Red Thermograms and Their Evaluation," Measurement and Automation XXVI, 5, 172-178 (1978).

8. Kreith, F., "Transmission de la chaleur et thermodinamique," Massonn et Cie, Editeurs (1967).

9. Bramson, M.A., Sprawoczuyje tablici po infrakrasmoe izluczeni ju nagrietych tiel, Moscow (1964).

10. Lloyd, J.M., *Thermal Imaging System*, Plenum Press, New York (1975).

11. Touloukian, Y.S. and D.P. DeWitt, *Typical Values of Emissivity*, Heyden and Sons, Ltd., Spectrum House, London,UK.

12. Hungarian patent pending.

Figure 3. Photograph of the Radiant Burner Containing Ceramic No. 3 with Treated Surface.

Figure 1. The Radiant Burner Containing Ceramic No. 2 with Chromium-oxide Treated Surface.

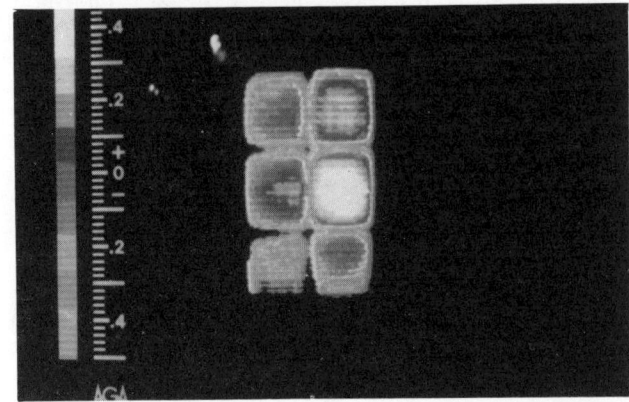

Figure 4. Thermogram of the Radiant Burner Containing Ceramic No. 2 with Chromium-oxide Treated Surface. $t_1 = 995 -$ °C; $t_2 = 925 - 995$ °C; $t_3 = 856 - 925$ °C; $t_4 = 791 - 856$ °C; $t_5 = 728 - 791$ °C; $t_6 = 660 - 728$ °C; $t_7 = 594 - 660$ °C; $t_8 = 540 - 594$ °C; $t_9 = 492 - 540$ °C; $t_{10} =$ $- 492$ °C.

Figure 2. Thermogram of the Radiant Burner Containing Untreated Ceramic No. 1. $t_1 = 892 -$ °C; $t_2 = 850 - 892$ °C; $t_3 = 814 - 850$ °C; $t_4 = 777 - 814$ °C; $t_5 = 743 - 777$ °C; $t_6 = 708 - 743$ °C; $t_7 = 675 - 708$ °C; $t_8 = 640 - 675$ °C; $t_9 = 608 - 640$ °C; $t_{10} =$ $- 608$ °C.

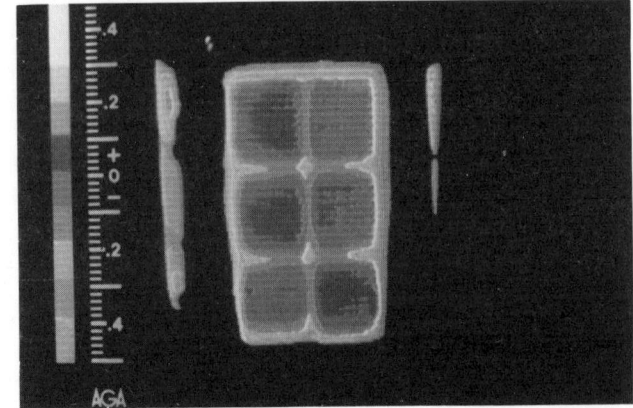

Figure 5. Thermogram of the Radiant Burner Containing Ceramic No. 3 with Treated Surface. $t_4 = 916 -$ °C; $t_5 = 840 - 916$ °C; $t_6 = 775 - 840$ °C; $t_7 = 713 - 775$ °C; $t_8 = 658 - 713$ °C; $t_9 = 606 - 658$ °C; $t_{10} =$ $- 606$ °C.

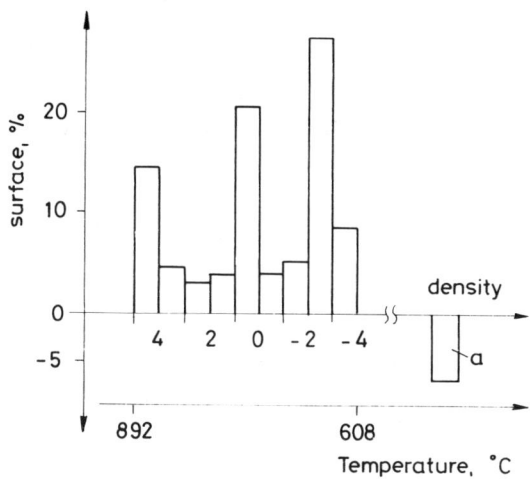

Figure 6. Density and Temperature Pattern on the Surface of Radiant Burner No. 1.

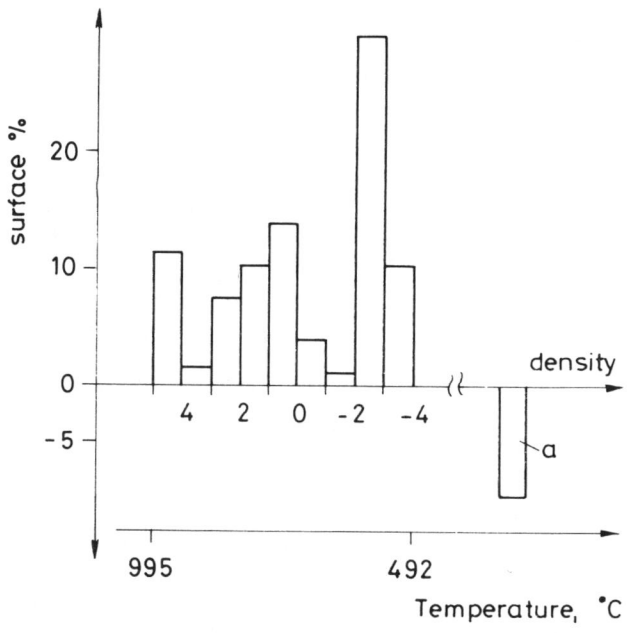

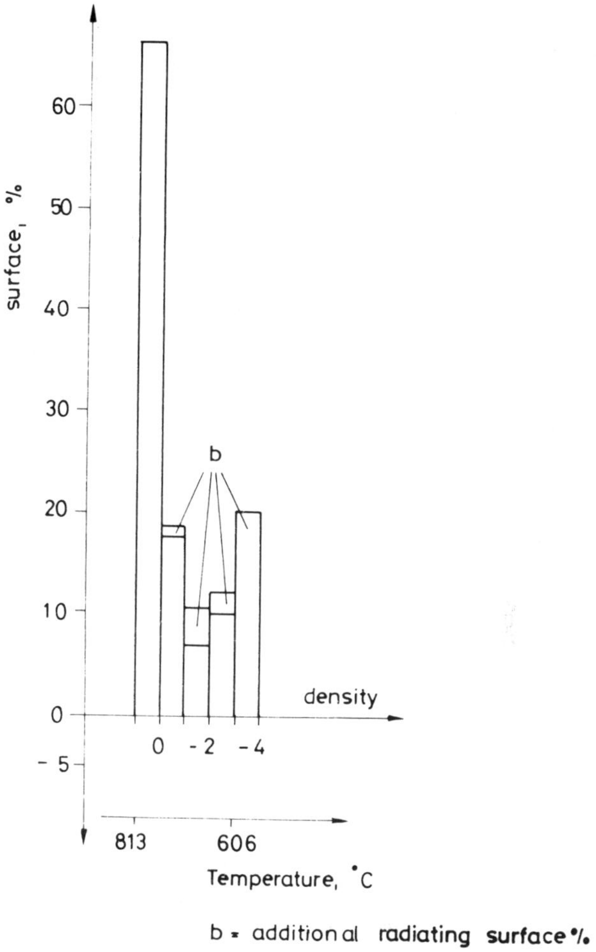

Figure 8. Density and Temperature Pattern on the Surface of Radiant Burner No. 3.

Figure 7. Density and Temperature Pattern on the Surface of Radiant Burner No. 2.

PROCESS CONTROL TUBE-SKIN THERMOCOUPLES

James G. Seebold ■ Chevron Corp., San Ramon, CA 94583-0945

In most process furnaces, the main problems are accuracy and reproducibility—and occasional failures. By contrast, in the higher-temperature pyrolysis furnaces, primary ammonia reformers and steam reformers, the real headache is getting the thermocouples to survive at all. Most rugged thermocouples assemblies produce temperatures higher than the actual tube metal temperatures. Type A thermocouples assemblies accurately reproduce the actual tube-skin temperature and last a long time.

INTRODUCTION

Thermocouples placed on the outside of furnace tubes warn against overheating. For a given operating pressure of the feedstock inside the tubes there is a maximum permissiable operating temperature of the tubes, and vice versa, dictated by the alloy and wall thickness of the tubes. Thus, there is a safe temperature-pressure envelope, within which the furnace must operate.

Operational requirements sometimes challenge that envelope. Perhaps we have to heat more feedstock, or the same amount to a higher outlet temperature or from a lower inlet temperature. Perhaps we have to change feedstocks, thus altering the thermal transport properties and process heating requirement. Perhaps coke deposits build up. On auto-fuel control valves to the burners open, the furnace starts firing harder, and the tube wall temperatures go up. Of course, we set alarms and limits to protect the furnace.

Inevitably, the questions then come up. Can we move those alarm set points up? How good are our tube-skin temperature measurements? How much conservatism was built in because they aren't perfect? Do we need all that? How imperfect are they? And so forth. Accurate tube-skin temperature measurement is important to the refining and chemical industry.

In principle, carefully assembled thermocouple junctions, which are carefully peened into carefully prepared notches in the tube surface, produce accurate tube-skin temperature measurements in the laboratory. The trouble is, they are not rugged enough for extended service in the actual process furnaces. The various approaches to "ruggedizing" the nearly-perfect laboratory thermocouple have complicated what is in principle simple. Almost all of the more rugged thermocouple assemblies produce temperatures that are variously and significantly higher than the actual tube metal temperature.

A tube-skin thermocouple assembly, commercially available, is rugged enough for service in refinery process furnaces and accurately reproduces the actual tube-skin temperature. But we do not know why it works and the other don't. This configuration would gain wider acceptance, and tube-skin temperature measurement would improve in the refining and chemical industry, if the heat transfer experts could explain why that's exactly what ought to happen.

Tube-skin thermocouple problems relate to longevity and placement. No matter how well designed or located, the thermocouple is useless as soon as it fails for whatever reason. Most thermocouple failures are really lead failures. The leads must be taken out of the furnace with the least possible exposure to radiation and flame. Often they are clipped to tubes to afford as much protection as possible in their route out of the combustion zone. The best path is determined by common sense.

There are two location problems. First, the hottest spot on the furnace tube can't be measured if there is no thermocouple there. Tube-skin thermocouples must be placed where the tubes are expected to be the hottest, based initially on process design condition and later on observations of the furnace in operation. Thermocouples should be retrofitted to locations that are observed to get hot if none were provided there in the initial installation. Infrared scanning techniques are useful in this respect.

Chevron requires at least three tube-skin thermocouples per pass, normally one or more on the outlet tubes as these are usually the hottest. A method of predicting furnace tube hot spots due to direct flame radiation is used. In vertical cylindrical furnaces with burners on the floor, for example, hot spots are usually found about one-third of the way up in the radiant section on the outlet tubes.

The second location problem is the circumferential temperature variation. The temperature on the back side of the tube that has no view of the flames is 50-150°F (28-83°C) lower than on the side facing the fire. It is poor practice to put the thermocouple on the back side of the tube away from the fire.

TYPE A THERMOCOUPLES

In 1935, a study reported that the old Socal Standard thermocouple, Figure 1, read 30-60°F (17-33°C) higher than actual. Thereafter, there was a great deal of interest in the swaged magnesia welding-pad type of tube-skin thermocouple, Figure 2. Through 1970, over a dozen different Chevron tests compared this type with the old Socal Standard. The collective results suggested that the welding-pad type read 40-100°F (22-56°C) higher than the Socal Standard. Still other Chevron comparisons indicated that welding pads read about 25-125°F (14-69°C) higher than actual.

The only firm conclusion drawn from these tests was that properly installed tube-skin ther-

mocouples of the two types in common use generally read high by 30-150°F (17-83°C). The Socal Standard had been used successfully for years. It *was* long lasting. Everybody knew it read high, but nobody knew how much. In 1975, Chevron ran comprehensive tests on a number of other promising designs, see Figures 3-6.

Thermocouples can generally be classed as unshielded, shielded, and shielded with a refractory-filled shield. The welding-pad type, Figure 2, is an example of the unshielded type, and the old Socal Standard, Figure 1, is an example of the shielded type. Based in part on the favorable experience of others, we judged the refractory-filled shielded type A, Figure 5, to be a promising candidate. Needed was a side-by-side comparison test with accurate, but fragile, peened thermocouples and other commercially available, rugged alternatives.

COMPARISON OF THERMOCOUPLES

A direct comparison test was carried out in an operating furnace. A variety of commercial thermocouples, interspersed with peened reference thermocouples, were arrayed in close proximity on a single tube. The test confirmed that the refractory-filled shielded type A gave accurate readings. The welding-pad type of thermocouple typically read 969°F (521°C) compared with its adjacent peened thermocouple reference temperature of 923°F (495°C), while the refractory-filled shielded type A read the same as its adjacent peened thermocouple reference temperature, 927°F (497°C).

Since standardizing on the refractory-filled shielded type A thermocouple assembly in 1978, field experience has been excellent. Both new construction and retrofit installations confirm the longevity of the type A thermocouples that has already been experienced by other major refiners. With older skin point designs, we occasionally said: "Don't worry about moderately high skin points. They always read high, anyway." *Type A, however, does NOT read high!*

Many type A thermocouple assemblies have been retrofitted into older furnaces, either replacing failed skin points or augmenting the existing pattern. All of our experience has been good, except for occasional *lead* failures. For example, in an older arbor-type catalytic reformer furnace,

100 type A assemblies were installed. At or shortly after startup, 20% "failed." This situation is not unusual in view of the difficulty of protecting the leads from the flames in this type of furnace.

In furnaces with wall-hung tubes, it is easy to extract the leads by running them around the tubes and out the back, away from the burner flames and directly through the furnace wall. But in a catalytic reformer furnace with tubes hung in the middle of the firebox, it is not so easy. In such a furnace, the best route out of the firebox is to clip the thermocouple leads to the radiant tubes and run them straight down and out the floor. We didn't do that initially. We learned the hard way.

The thermocouple assemblies discussed are all of the traditional chromel-alumel material, which are good for low to moderate tube temperatures. More recently, we have found that NiSil-NiCroSil assemblies have good staying power in several of our steam reforming furnaces, in which the costly alloy tubes typically run at 1,700-1,800 °F (927-982 °C).

We do not have extensive service experience yet. But thus far, discounting infant mortality, we are getting temperature indications that are close to optical pyrometer measurements and service lives in excess of one year, which is about a year longer than we've ever gotten before. We intend to continue testing these promising high-temperature thermocouple assemblies.

SUMMARY

Test experience since 1935 with the old Socal Standard and other thermocouple assemblies in common used showed that properly installed tube-skin thermocouples generally read high by up to 150 °F (83 °C). On the other hand, they were often placed on the back of the tube, away from the fire, where it is cooler by up to 150 °F (83 °C). To some degree, these errors balanced one another, but nobody knew how much.

Type A thermocouple assemblies, properly placed on the side of the tube that faces the fire, accurately reproduce the actual tube-skin temperature and they last a long time. Today, Chevron regards type A as the best answer to furnace tube-skin temperature measurement in most process furnaces.

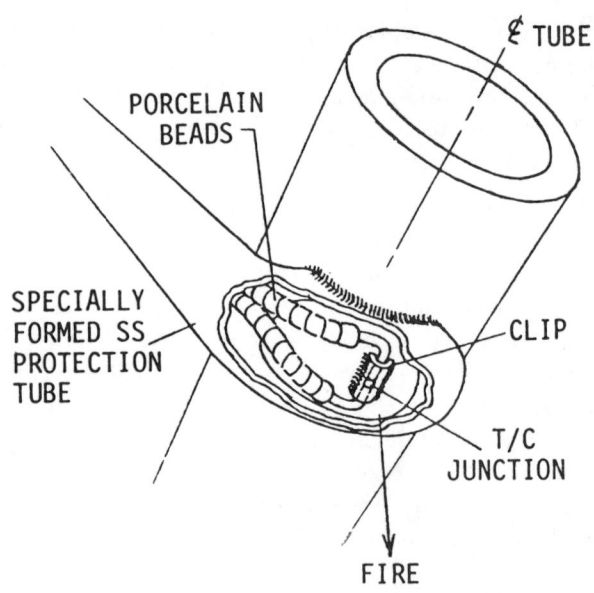

Figure 1. Socal "Standard" (Shielded) Thermocouple.

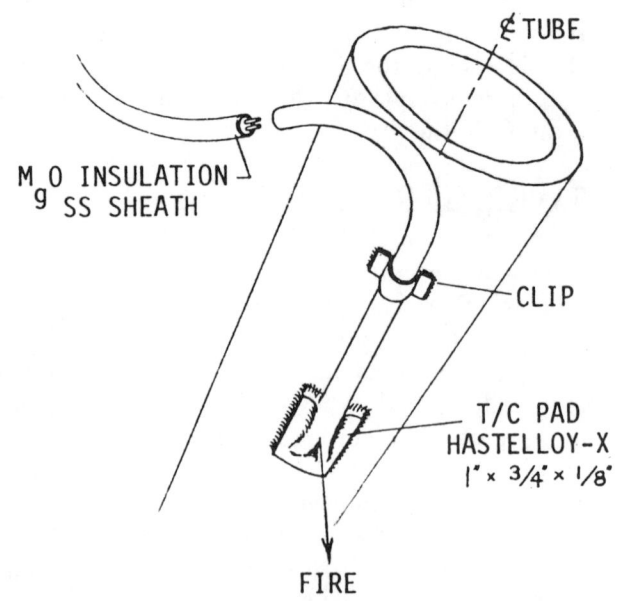

SI Conversion: mm = in. X 25.4

Figure 2. Welding-Pad (Unshielded) Thermocouple.

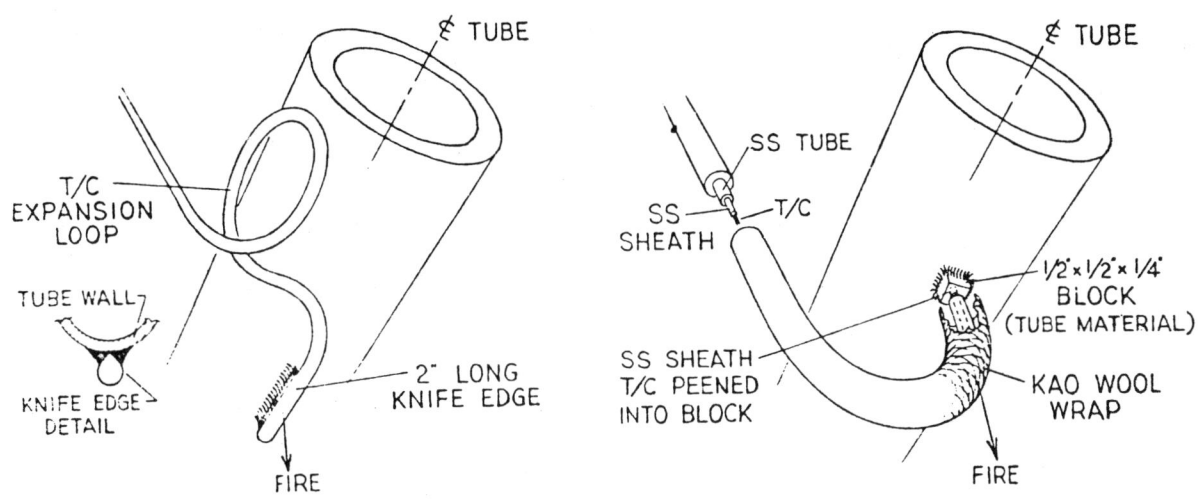

Figure 3. Other Unshielded Type Thermocouples.

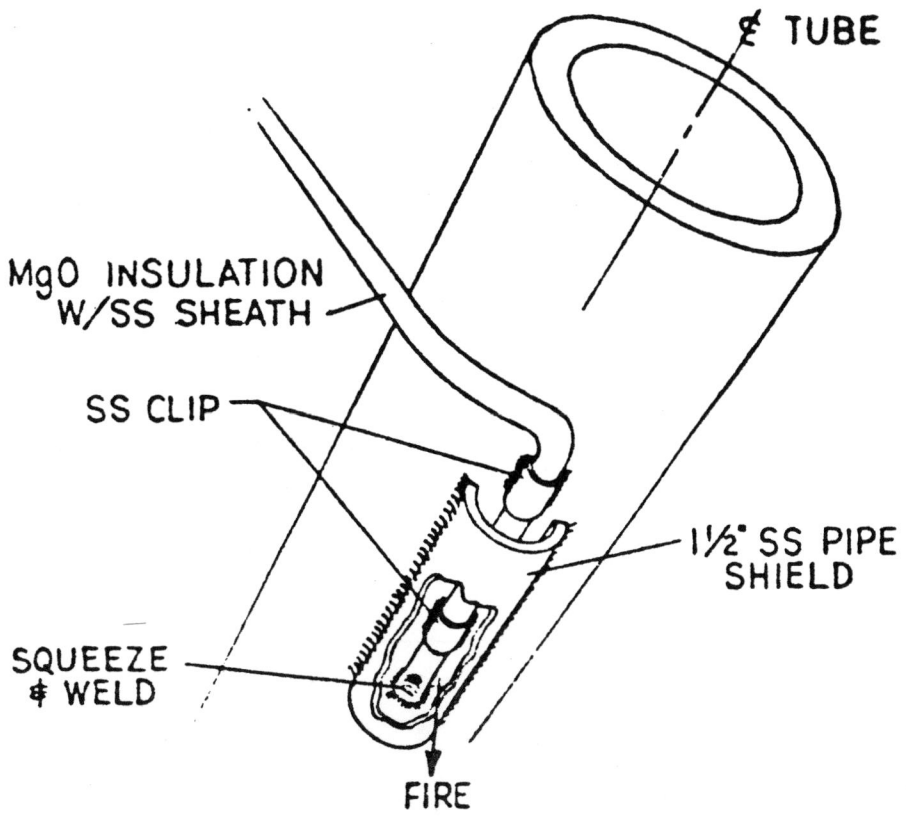

Figure 4. Another Shielded Type Thermocouple.

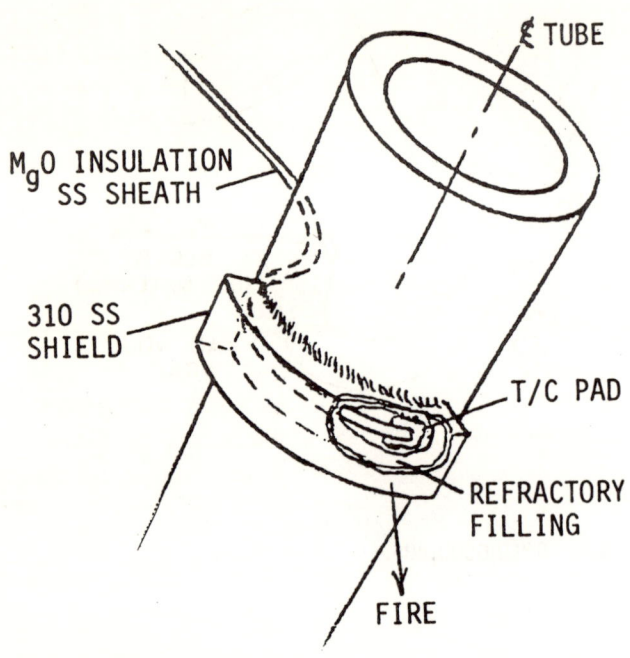

Figure 5. Refractory-Filled Shielded Type "A" Thermocouple.

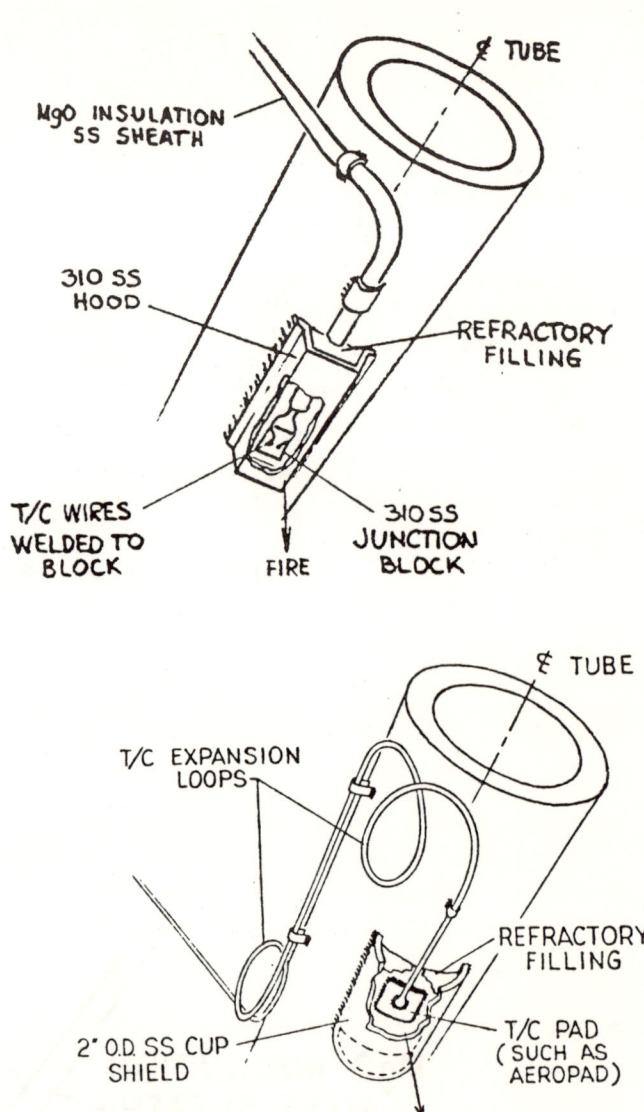

Figure 6. Other Refractory-Filled Shielded Thermocouples.

MEASUREMENT OF TUBE-SKIN TEMPERATURES OF PROCESS FIRED HEATERS

H. Sugano, T. Hamazaki and M. Ise ■ JGC Corporation
14-1, Bessho 1-Chome, Minami-Ku, Yokohama, 232 Japan

Thermocouples were used to measure tube-skin temperatures. Several modifications were tested including sheathed thermocouples, use of welding pads or hoods, and knife-edge thermocouples. In general, the temperatures indicated by the thermocouples are higher than the true temperatures. The results reported here will aid in selecting and installing thermocouples.

INTRODUCTION

Process fired heaters frequently serve as heat sources in petroleum refining and petrochemical process plants. Pressurized flammable hydrocarbons in such heaters often flow through the tubes heated directly by flames in fire boxes. Monitoring the tube-skin temperature is highly important for safe operation in order to maintain the tube materials within their design limitations. Especially for processing of heavy oils, coke frequently forms inside tube walls resulting in higher tube-skin temperatures; monitoring skin temperatures is especially essential in such cases in order to control the allowable throughput and to determine when to decoke.

In general, the skin temperatures of a tube are measured by contacting the metal surface or skin with thermocouples, and several types have been adopted (1,2). Overall accuracy of the measurement depends on the method of attaching the thermocouples to the tube surface. Little technical data have, however, as yet been reported indicating how well the thermocouple readings represent the true tube-skin temperatures. Tube-skin temperatures can be calculated using API Recommended Practices (3).

In the present investigations, several types of tube-skin thermocouples were installed in an experimental fired heater, and temperature differences were determined for several types of thermocouples for comparable temperatures.

TYPE OF TUBE-SKIN THERMOCOUPLES

Types (A) to (G) as shown in Figure 1 are skin-thermocouples that are frequently used industrially, and they were investigated in the current investigation. Types (A) to (F) are sheathed thermocouples; type (G) is, however, a bare thermocouple element. Types (A) and (E) are most commonly used (1,2). The thermocouple is normally connected to the tube surface by welding; direct welding may, however, damage the thin sheath. A welding pad or tip is, therefore, usually attached to the thermocouple of types (A),(B),(D),(E), and (F). Type (C) known as a knife-edge is also sheathed (4). The sheath of this type is relatively thick permitting direct welding and better service life, and no welding pad is needed.

Some thermocouples are covered by a hood or enclosed within a tube. Such a protection tube or a cover is necessary for bare thermocouples such as type (G) in order to isolate the elements from high-temperature and corrosive combustion flue gases. The cover or tube as shown in types

(B) and (F) prevents excessive local heat absorption that causes errors in temperature readings.

Type (H) was developed for the present investigation. The thermocouple element is basically the same as type (E). However, a shielding cover is provided to prevent excessive heat flow into and through the sheath. The cover was spot-welded on the tube in order to minimize the heat flow into the tube through the cover.

APPARATUS AND PROCEDURE

Figure 2 shows a schematic drawing of the experimental apparatus. The fired heater employed is a vertical-cylindrical type with 2-inch nominal diameter tubes as described in Table 1, and air was introduced as a coolant into the tubes using a blower. The details for the thermocouple installation are listed in Table 2.

In addition to the eight types of thermocouples [types (A) to (H) in Figure 1], two thermocouples were installed as standards in order to measure "true" tube-skin temperatures, see Figure 3. These two thermocouples had thin sheaths and were embedded in the tube walls; such an installation does not disturb the heat-transfer conditions.

Temperature readings of the tube-skin thermocouples, flow rates and temperatures of the cooling air, and temperatures of combustion flue gas were simultaneously recorded. Several different heater operating loads were investigated by controlling the burner firing rate.

RESULTS

The temperature readings are reported for thermocouples, types (A) to (H) in Figures 4 through 11. Temperature differences between the tube-skin and the air flowing in the heater tube were correlated as a function of the average heat fluxes as shown in these figures. The average heat flux is defined as the heat absorbed through a unit tube outside surface area per unit time, and the heat absorption rates were calculated based on the flow rates and the temperatures measured. Figure 12 shows the comparisons of testing thermocouple data with standard thermocouple data and calculated results. The tube-skin temperature was calculated using the API Recommended Practices (3) which is generally employed in the design of process fired heaters.

The results are summarized as follows:

(1)   The temperature readings vary widely for the various thermocouples tested, as indicated by comparing Figures 4 through 11. The maximum difference which is observed between type (A) and type (H) range from about 85 to 90° C, see Figures 4 and 11.

(2)   All types of thermocouples tested indicated higher temperatures than both the "true" tube-skin temperatures measured by the embedded thermocouples and the calculated temperatures. Types (A) and (E) indicate comparable readings that are much higher than those of the other thermocouples, see Figures 4 and 8. Type (C) and (D) thermocouples and also thermocouples with a protective tube or cover such as types (B), (F) and (H) resulted in relatively low temperature readings; however, the deviations from the "true" temperature are still in a range of 30 to 40° C, as indicated by Figure 12.

DISCUSSION

In addition to the basic thermocouple accuracy and instrument error, the following factors contribute to deviations from the "true" temperature.

(1)   Incomplete contact of thermocouple element with tube surface.

(2)   Heat input through thermocouple assembly.

(3)   Heat shielded by protective tube or cover.

Factors (1) and (2) increase the apparent temperature but factor (3) lowers it. Factors (1) and (2) seem to be highly important for types (A) and (E) thermocouples. The sheathed thermocouple element of type (A) is fixed on the welding pad (25 mm square and 3 mm thick stainless steel pad) so that the temperature sensing point does not contact directly with the tube surface; and an gas gap between the pad and the tube surface is inevitable. The gas gap, even if it is very thin, will have an important influence on the reading since the thermocouple assembly is exposed to the high-temperature flue gases in the fire box. For type (E), the sheathed thermocouple wire protrudes into the high temperature surroundings, and seems to conduct excessive local heat to the skin point. Comparing type (E) with type

(H), the heat shielding is effective for minimizing excessive heat flow through thermocouple assembly. However, it should be noted that the protective cover or tube, if it is installed on the tube surface as in the case of type (H), should be small in order to minimize excessive radiant-heat shielding on the measuring point of tube surface.

SUMMARY

Temperatures as measured by thermocouples attached to, or contacted with, the outer wall of the tubes in process-fired heaters are, in general, higher than the true wall temperatures. The results of the present investigation will aid in selecting and installing thermocouples to make temperature measurements since a knowledge of the true temperatures are highly critical in the design and operation of a furnace. The findings of this investigation indicate the need for development of better methods of measuring temperatures in furnaces.

REFERENCES CITED

1. API Recommended Practice 550, "Manual on Installation of Refinery Instruments and Control Systems," Part I "Process Instrumentation and Control," Section 3 "Temperature," 3rd edition, 1976.

2. ibid., Part III "Fired Heaters and Inert Gas Generators," 2nd edition, 1977.

3. API Recommended Practice 530, "Recommended Practice for Calculation of Heater Tube Thickness in Petroleum Refineries," 2nd edition, May 1978.

4. US Patents, No. 3874239 and 3939554.

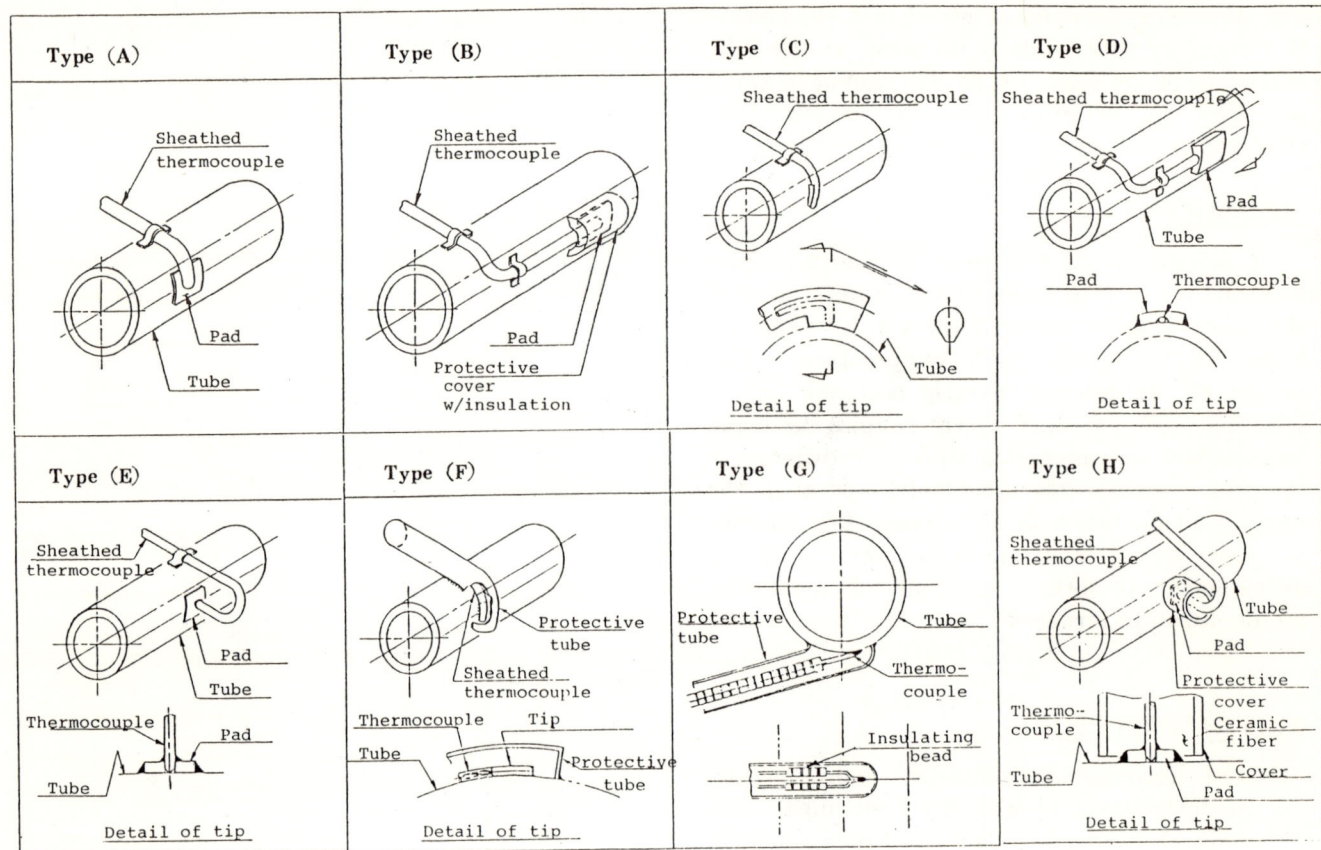

Figure 1. Types of tube-skin thermocouples.

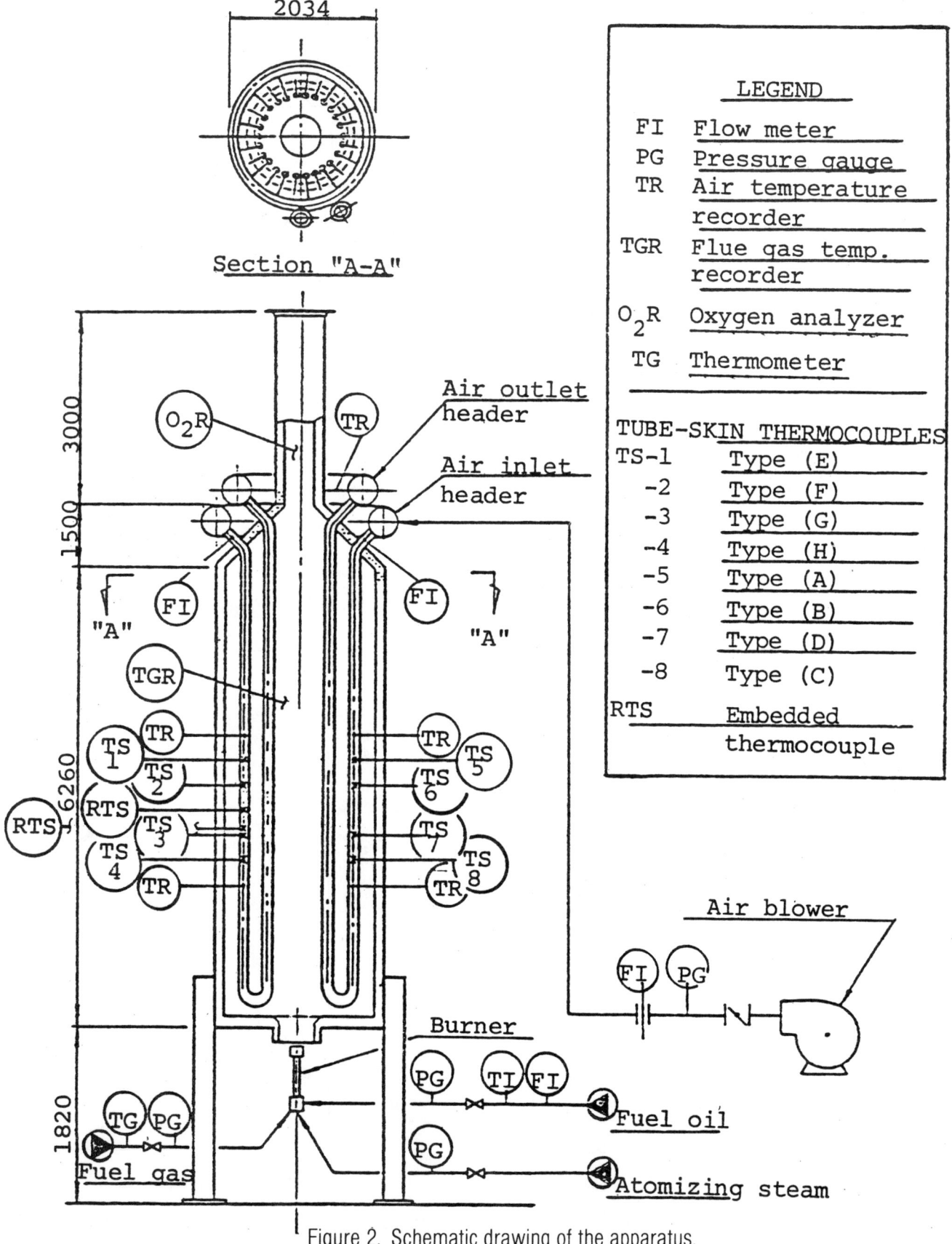

Figure 2. Schematic drawing of the apparatus.

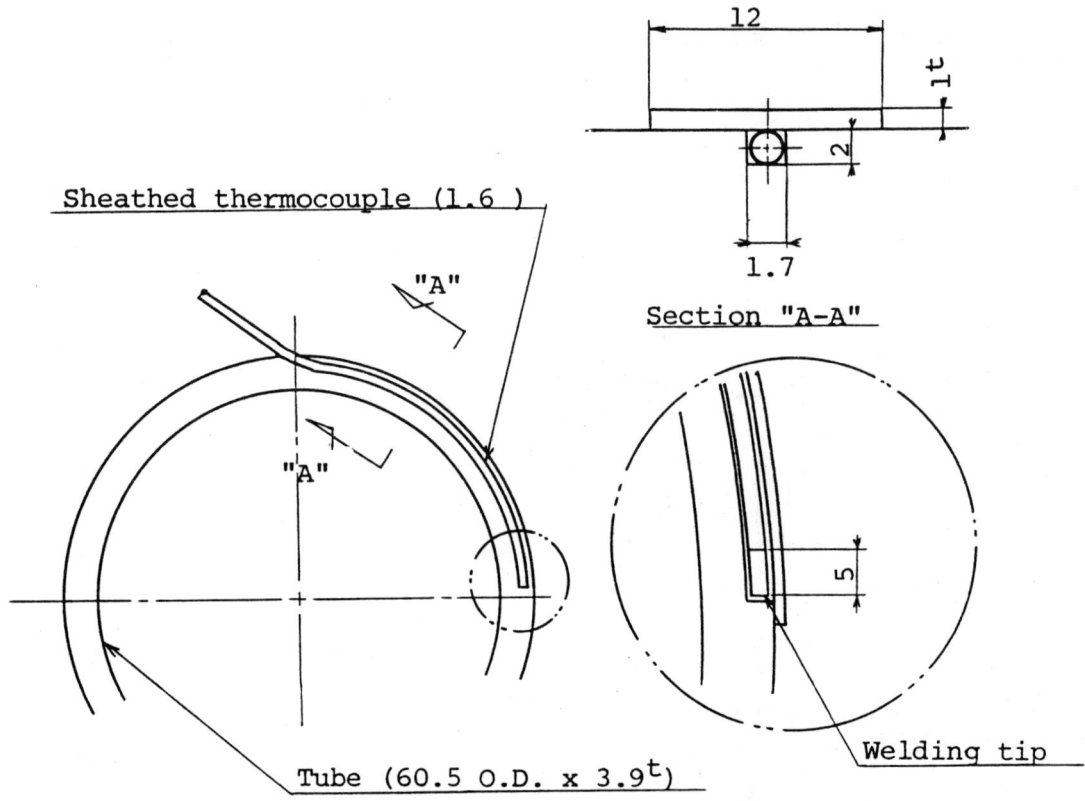

Figure 3. Thermocouple embedded in the tube walls.

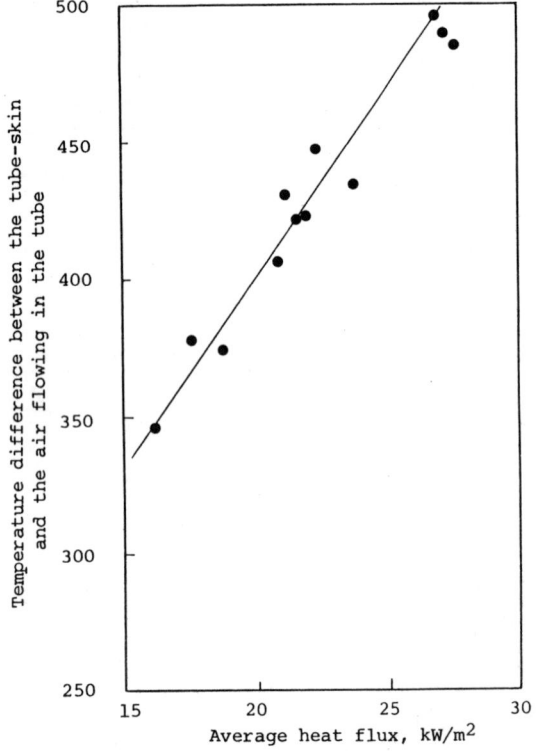

Figure 4. Temperature indication of Type (A).

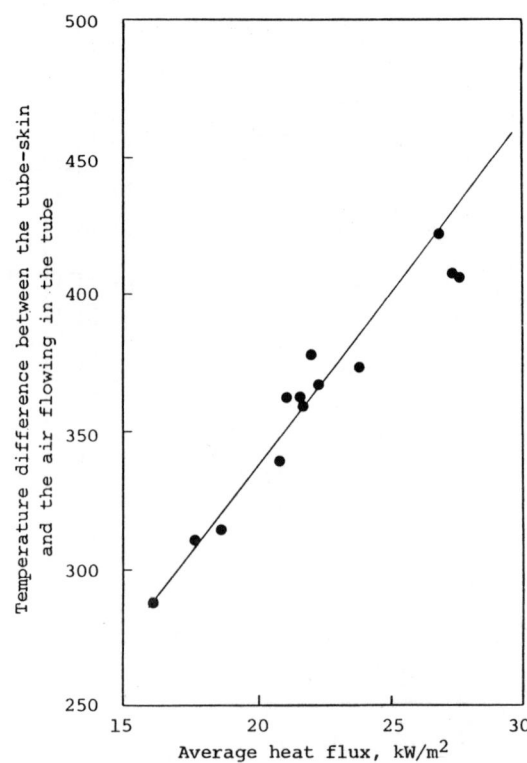

Figure 5. Temperature indication of Type (B).

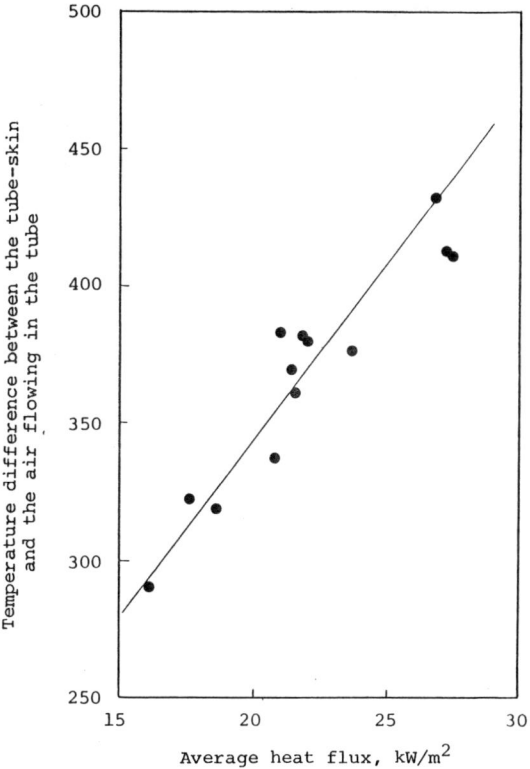

Figure 6. Temperature indication of Type (C).

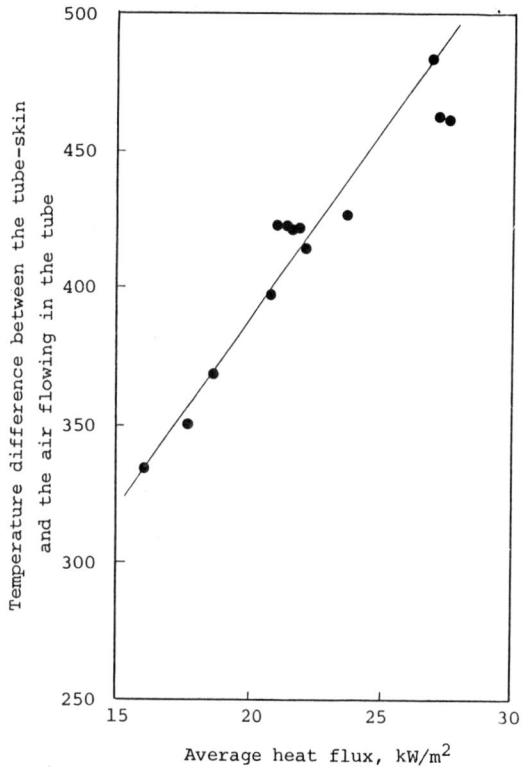

Figure 8. Temperature indication of Type (E).

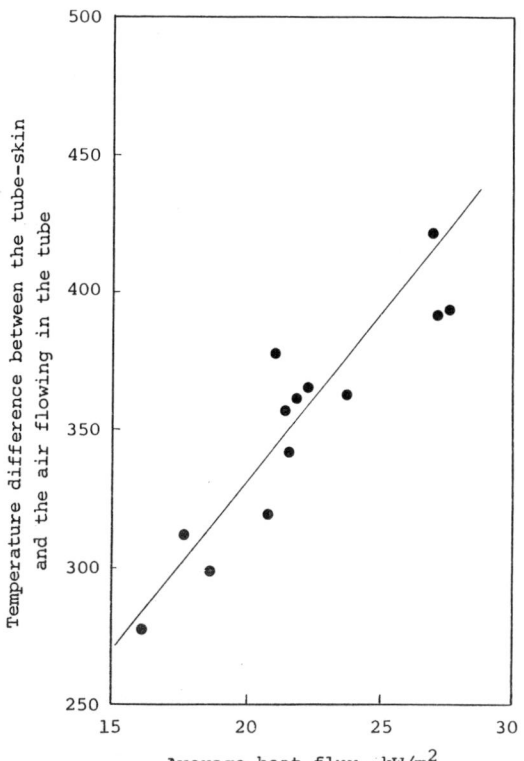

Figure 7. Temperature indication of Type (D).

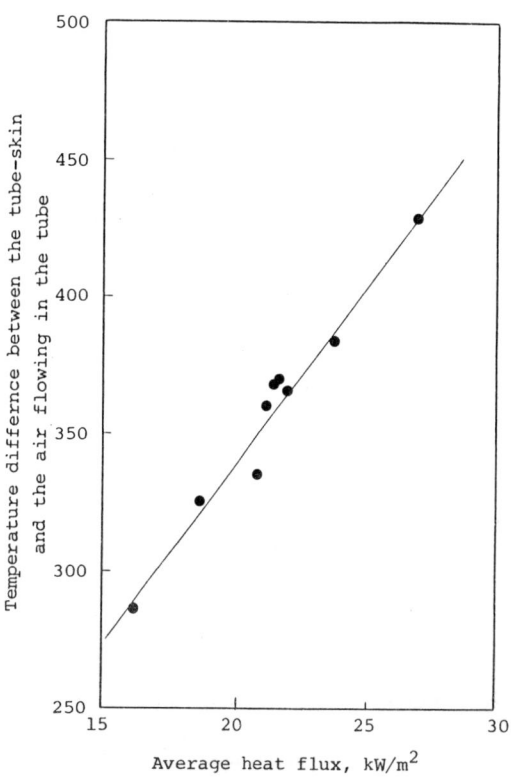

Figure 9. Temperature indication of Type (D).

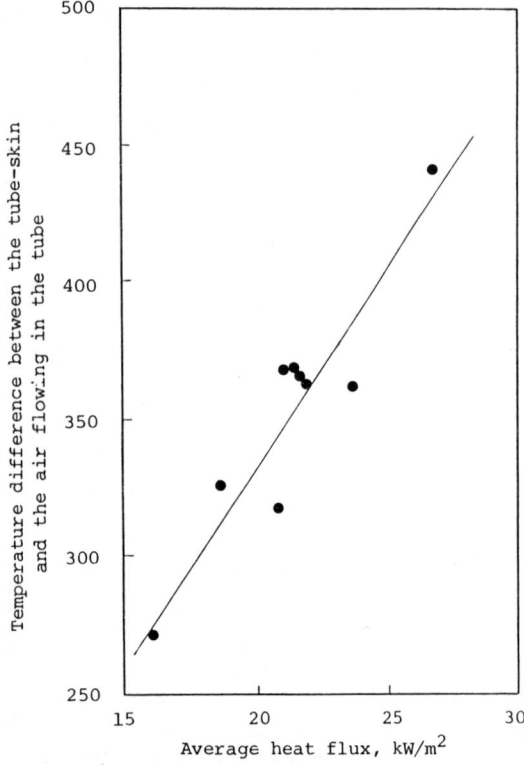

Figure 10. Temperature indication of Type (G).

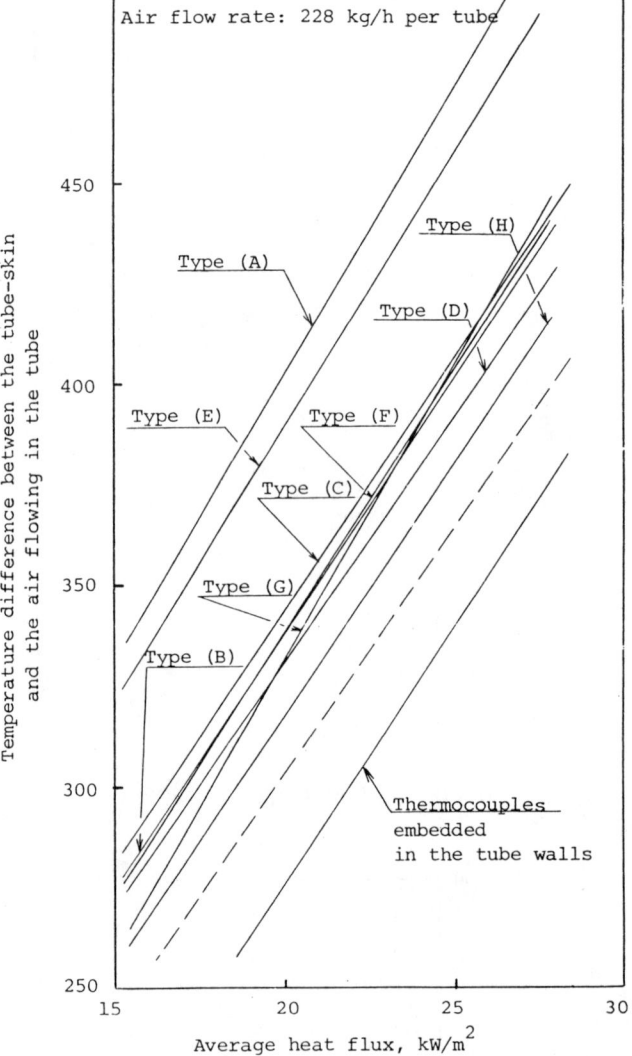

Figure 12. Comparison of temperature indications.

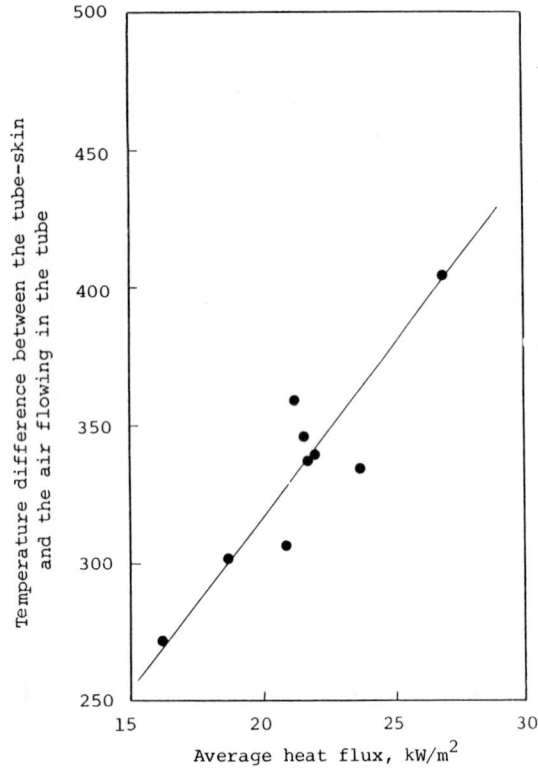

Figure 11. Temperature indication of Type (H).

Table 1   Specification of the heater

(1)	Type	Vertical-cylindrical
(2)	Dimensions	Approx. 1.8 m x 6 m high
(3)	Heating tubes	
	(a) Size	60.5 mm O.D. x 3.9 mm t
	(b) Material	2-1/4 Cr - 1 Mo (JIS STPA 24)
	(c) Number of tubes	26
	(d) Number of streams	13
	(e) Arrangement	Single row against wall
	(f) Tube center to center distance	197 mm
(4)	Burner capacity	2.3 MW on LHV basis

Table 2   Tube-skin thermocouples employed in the experiment

Type	Element (Sheath: Type 316S.S.)	Tip or Pad (Type 304S.S.)	Protective Cover (Type 304S.S.)	Thermal Insulation
(A)	8 sheathed thermocouple (Chromel-Alumel)	25×3^t	-	-
(B)	ditto	ditto	$40^W \times 45^L \times 35^H$ (2^t)	Packed w/ceramic fiber
(C)	ditto	-	-	-
(D)	ditto	25×12^t	-	-
(E)	ditto	25×3^t	-	-
(F)	ditto	8×15^L	$34^{O.D.}$ tube	Packed w/ceramic fiber
(G)	3.2 bare thermocouple (Chromel-Alumel)	-	ditto	-
(H)	8 sheathed thermocouple (Chromel-Alumel)	14×3^t	$34^{O.D.} \times 50^L$ tube	Packed w/ceramic fiber

"Note": (1) Type of junction: Grounded for all the thermocouple elements
(2) All dimensions in mm

PANEL DISCUSSION: THE TECHNOLOGY OF HIGH-TEMPERATURE MEASUREMENTS

Panelists ■ 1. D.P. DeWitt (Moderator), Purdue University, West Lafayette, IN
2. W.R. Barron, Williamson Corporation, Concord, MA
3. D.G. Damin, E.I. duPont de Nemours and Co., Inc., Beaumont, TX
4. H.C. Hottel, Massachusetts Institute of Technology, Cambridge, MA
5. P.D. Hunsucker, Shell Development Co., Houston, TX

D.P. DeWitt, Purdue University - Moderator

Let me begin by introducing the panelists. This is in the order in which you will hear from them. First is Dean Damin who is a metallurgist and a materials expert at DuPont's Beaumont Operations. Second is Paul Hunsucker, who is in the Instrumentation Division of Shell Development Co. Next is Bill Barron from Williamson Corporation, who represents the instrument manufacturers on the panel. And, finally, is Prof. Hottel from the Massachusetts Institute of Technology (MIT). Dr. Hottel, you without question represent the fundamentals of thermal radiation, and related subjects from the 1930's to the present time.

The purpose of this panel is to provide an opportunity for an exchange of ideas. Since no similar symposium has been held for many years on petrochemical applications, such an objective seems most appropriate. We hope to identify current and future requirements for measurement technology. In the instructions to the panelists, five points were suggested as being appropriate:

1. The importance of accurate temperature measurements to the petrochemical industry. Is it important to do better than is being currently done?

2. The assessment of the measurement technology using thermocouples and radiation thermometry. How would you judge the technology today? Adequate? Is it going to improve normally to meet future requirements, or will other things be required?

3. The role of the instrument manufacturer, the technical societies, or other organizations in solving measurement problems and in disseminating information.

4. Personal experiences concerning any aspect of measurements with thermocouples or radiation thermometers.

5. General and specific reactions of the panelists to the papers presented earlier in the symposium.

Our format is as follows: Each panelist has prepared approximately ten minutes of comments on any and all of these topics; following their presentations, questions can be asked from the floor. Following this, other people will hopefully respond on matters they consider important.

Our first panelist is Dean Damin from DuPont.

D.G. Damin, DuPont - Panelist

I've been asked to make some specific comments about the papers of the last day and a half and what I think is the status of temperature measurements. Before I tell you what I think, I will take just a moment to tell you about my platform and my background. For 10 years prior to June 1984, I worked as a metallurgist in the materials and corrosion group of Gulf Research and Development Company, in Pittsburgh, PA. That is where I obtained most of my experience. As yet, I have done little materials-oriented work for DuPont, although I hope to get involved in it. I am a metallurgist and I do not understand how the engineers employ all their formulas and get all these great numbers.

I am concerned about materials of construction, and how they hold up in service. Temperature measurements are vital in the life of furnace tubes. If a heater is designed for a given temperature, typically the tubes are designed to last 10 years at that design temperature. If the temperature of operation is increased $100\,^\circ$F, the 10 years life suddenly drops to one year of life. It does not require much increase of temperature to adversely affect tube life. As a metallurgist, I am asked what is the optimum tube life? Where does the unit obtain the most economical, the most optimum tube life? I need to know the temperature--it is important. I also receive questions when the plant is operating at 120% of capacity because that amount can be sold. I am asked what happens by running hotter and harder. Again, it is important to know how hard, not from a percent of design, but from a temperature standpoint. It is also important from a safety point and based on maintenance planning.

Short tube life does not necessarily mean it is less than optimum. The economics may indicate that one year at this high severity is wonderful since more than adequate money is made to replace those tubes. One needs to know however that the tubes will last at least a year, so one can plan for it and so one can run safely. So that is my background — I am a metallurgist in this world of tube wall temperature measurements.

At his symposium, there has been an excellent mix of academic, instrument manufacturer, and instrument user papers. We have gone from the basics to the extremely practical. I have heard many interesting points. Some I guess I knew before, but they just never really clicked. I really did not **know** them before. I suspect I am not the only one in the audience who has gone through that process. Probably we all learned a lot, but even if I am the only one who did learn anything, I can guarantee there are many people in operating plants who need to hear what has gone on in this room the last day and a half.

Next, where do we go from here? I just say "Let's continue." Lets continue the interaction between the instrument manufacturers, the designers, and the instrument users. Let's continue the strong *technical,* and I emphasize technical, interaction between these groups, supplemented by input from the academic world. There is certainly a strong economic incentive for both the instrument manufacturers and the instrument users for good temperature measurements. Manufacturers, you will sell instruments, it is as simple as that, that is how you will make a profit. Relative to the process industries when you get a much better handle on tube metal temperature, you too will realize significant increased profitability. The instrument manufacturers contribute considerable to the physics backgrounds that, basically, the operations people in the plant do not have. Operations personnel can however contribute; they know what is in the fire box. They do not always tell everything, maybe because they do not think it is important, maybe because we have forgotten. With the right questions going back and forth, plant personnel should be able to get the maximum use of the available equipment and knowledge that is now available.

My last comment is if there is one group that seems to be missing, it is the *furnace* designers and manufacturers. Based on my personal experience, those people have a poor handle on actual tube metal temperature measurement and what is actually going on in their furnaces. Concerning my questions on tube metal temperatures, the answer always begins and ends with a long equation as to what the calculated temperature will be, but never much help as what it actually is.

G.R. Peacock, Land Instruments: Addressing the comment about the interaction between the users and manufacturers, there is a vehicle in place, the ASTM Committee on Radiation Thermometry (E20.02). We encourage companies to send representatives or have representatives on our mailing lists. Companies should be aware we are developing standards and educational material for the application of radiation thermometry in numerous industrial applications.

D.P. DeWitt: That particular committee appears to be well represented by the manufacturers. But there are no representatives from the petrochemical industry, which is unfortunate and a real void.

D.G. Damin: I wish to make another reference which goes back to an opening comment by Prof. Albright yesterday. The Materials Technology Institute of the Chemical Process Industries sponsored a program through Drs. DeWitt and Albright, to assemble a report of the state-of-the-art technology. So that is another step, but we must continue making these steps forward.

J.W. Friesell, Infrared Scanning Services: We make many measurements of tube temperatures inside fire boxes. It is good talking to a metallurgist because I hear everybody say "Wow, 100 ° F, and I am going to lose my tubes next week". How certain are we of the data that predicts the life of the tubes? People who are concerned and work hard to get temperature data as good as possible using the available technology are asking such questions.

D.G. Damin: There is a fair amount of creep and stress rupture data that was generated in laboratories under ideal environments. We are not talking about high severity furnaces, but run-of-the-mill furnaces with longer life, such as 100,000 hour design life. I know of tubes that have been in service 250,000 hours in some furnaces. Laboratory data needs to be extrapolated; one usually extrapolates after 10,000 hours and generally on the safe/conservative side.

T.G.R. Beynon, Land Infrared, Ltd. I am interested in Mr. Damin's final comment about

furnace manufacturers; as reported in the talk that I gave yesterday, my experience is mainly with furnaces in the steel industry. We have often thought that if the problems of infrared temperature measurement were considered during design of the furnace that one could easily make improvements over the present situation. The considerations of geometry, which are almost irrelevant to the furnace designers in terms of the operability of the furnace, could often make all the difference to infrared temperature measurements. Are similar features important in tube furnaces? Perhaps more consideration of infrared temperature measurements during the furnace design phase would result in a net gain improvement. One would likely have to compromise some cost aspects of design in order to permit easy measurement, but a significant gain in profitability might occur.

D.G. Damin: Anyone who has hung off the side of a scaffolding and has looked through a peephole at a tube on the other side of the furnace—because that is the important tube and this opening is as close as you can get to it—will agree with you.

D.P. DeWitt: Will a representative of a furnace manufacturer comment?

Representative from M.W. Kellogg Co.: Considering furnace designs and trade off costs versus operability, the first thing that suffers is accessibility to measuring temperatures. As we become more aware however of the need to measure the temperatures at the hottest spot, i.e. the critical location, improved access for temperature measurements will be provided. Peep sites are now available, at least on our furnace designs.

J.G. Seebold, Chevron Corporation: Design stage accommodations are a great idea, but we do not expect to build many furnaces in this country in the next 10 years.

J.W. Friesell: As a suggestion, an infrared imaging system can help locate the hottest spots in a furnace. We are now doing studies on some major crude heaters. Thermal mapping locates the hottest places inside the fire boxes so that a customer can locate thermocouples in the proper positions.

Representative from M.W. Kellogg Co.: The furnaces to which I am referring are often about 25-30 years old. At that time, the computer models for determining the location of the tubes with maximum temperature were not good. In the last ten years, modelling has improved tremendously, so we can now say, "This will be the location of the maximum temperature" and access can be provided to measure the temperatures at that point.

P.D. Hunsucker, Shell Development Co. - Panelist

At Shell Development we have tried to serve as a focal point for iron plants, chemical plants, and refineries relative to temperature measurements. In the past, we relied on instrument manufacturers for information on measuring temperatures. Unfortunately, we did not always contact the proper instrument manufacturers. In some cases, we had problems. For about the past 12 years, we at Shell have had programs to evaluate methods of measuring temperatures for pyrolysis furnaces with tube temperatures of 1600-1900°F. In our research and development effort, we have tried many things that did not work. Based on my thermocouple experiences, I realized how hot it is and how miserable it is making these measurements. We have tried thermocouples where we would take two very pointed wires with a supporting sheath and touch them to the tube. Sometimes we got a reading but generally we did not. We have tried several configurations including one compensated with water; a furnace manufacturer however recommends against water entering the furnace. We have also tried air cooling or compensating with air.

We have however now developed a calibration procedure for purchased pyrometers. A target or probe was built and it is positioned rather easily in the furnace. This target is equipped with replaceable thermocouples. With this method,

we can to obtain what are thought to be true surface temperatures for the tubes. The target that was developed is mounted on a rather small handle (actually a rather small diameter tube) so that the target can be inserted through a peephole and positioned in a furnace. The thermocouple leads are positioned inside the small diameter tube. Cooling air also enters through the small tube and this air is used to adjust the temperature of the target to approximately the same temperature as the process tubes inside the furnace. It is important to have the target in the same temperature range as the process tubes when calibrating the pyrometers. There is always doubt for the calibration of the pyrometers if target is 400-500°F hotter than the process tubes. So we often cooled the target to about 1700°F in order to be at approximately the same temperature as the adjacent tubes. With this cooling air, two thermocouples are required in the target in order to obtain the surface temperature.

Before discussing further our measurement technique, the pyrolysis furnaces will be described. They are rows of vertical tubes connected by U-bends. The oil enters the inlet to a tube and goes up and down in succession through three zones; getting continuously hotter in each zone. Viewing ports are provided in the bottom of the furnace. There are rows of viewing ports along about ten passes. The probe is positioned between the passes so that it is in a geometrical position that would represent a tube. Its position is varied up and down in the furnace between the tubes. We look at it through another port with our pyrometer. The glass is removed from the port because there is anywhere from 10-100% attenuation depending on how dirty the glass is. With this approach, we have found good representation of the true tube surface temperature. The target is a rather short piece of HP-45 furnace tubing being the same material as used in a furnace tube. A couple of thermocouples enter the target tube and are inserted on the wall of the target tube. Two holes are drilled into the tube wall targets. Type K sheathed (1/16" dia.) thermocouples are positioned in the holes. One is positioned one third of the way from the viewing surface and the other one third of the way from the back wall. The two readings are extrapolated to get accurate skin temperatures. This approach is more accurate as compared with all the other known methods. We use this target in our furnace to calibrate any pyrometer desired.

Concerning the choice of pyrometer, we have tested some pyrometers in our furnace to calculate or determine geometrical correction factors. After measurements are made in a given furnace, we would like to go to our process computer and plug in a correction factor. Then the operators would take their on-site readings with a pyrometer, plug the data into the computer, and get the corrected temperatures. We would also have it measure the wall temperature of the target. The only thing that is really new is a correction factor that we called a geometric correction factor that depends on the location in the furnace being measured. The emissivity of the wall has been considered but does not seem to help. The reflectivity of the tube or target has been investigated. Some of the data that we have taken has helped, but not to a point where we can say it works. Perhaps much more extensive data would prove something.

We have looked at five positions in the furnace; there are 40 tubes for the ten passes of the furnace. We calculate a geometric factor to correct the measured temperature to the true temperature as determined with our probe. Two different pyrometers were employed. One was an optical pyrometer with an optical passband at 0.65μm and another with an infrared radiometer at 0.95μm wavelength. Two different firing conditions were investigated for a given furnace: one was at a very high load and the other was at a more moderate load. We corrected these results with our geometrical factors and obtained some improvement, but still not to the extend desired. Eventually maybe we can plug it into the computer and have the operator go out and take readings and do us much good. Perhaps on a given furnace with more data, the furnace could be characterized. One of our biggest problems found in measuring wall temperatures is that variations occur by as much as 200-300 °F over a relatively small distance. A 100 °F difference in reading has been noted on a tube one-third meter from the furnace wall; yet to the eye there did not seem to be the much temperature difference.

D.P. DeWitt: What values of emissivity do you use for your tube materials?

P.D. Hunsucker: For HP-45, the value is 0.85 to 0.90. For our measurements, however, the pyrometer was set at 1.0. This was done since the readings are always high, as indicated earlier, because of the reflection error. If the emissivity is set at 1.0, the readings are generally kept on scale.

D.P. DeWitt: It is then apparently good practice to defeat any emissivity compensation feature and to make corrections analytically.

P.D. Hunsucker: Until we know a lot more about what we are doing, yes.

E.M. Emery, E^2T Technology: We are an instrument manufacturer, and do not have a furnace. The best I get to do is to play with somebody's furnace for a couple of days. Your program obviously has taken months to accomplish; at least I am guessing it took quite a while. Are there any furnace operators that have the time, patience, and money to let instrument manufacturers come in at their own expense and work? The operators would run the furnace. If I had the facilities in which to work, I could do much more for in-furnace measurements than I have done. As it is, I set up tubes with torches, and try to make simulations in fire brick in the corner of my laboratory, but this equipment does not really simulate a plant unit.

P.D. Hunsucker: I wish I knew. I can not answer for Shell. It would be of interest to do something like that, especially in an area where operations have some latitude to adjust the furnace. If some changes are made in the furnace, safety must always be considered.

J.M. Soh, Arco Chemical: How often to you calibrate your pyrometers? You say you calibrate against a thermocouple, but do you calibrate the pyrometer itself against blackbody radiation? Why did you select the wavelengths 0.65 or 0.9 micrometers?

P.D. Hunsucker: At one of our plants, they have a calibration procedure employing a lamp in a box. Their pyrometer is calibrated at least once a week or whenever the pyrometer is dropped. The emissivity used is based on values from the literature. The emissivity of a sandcast tube is considered to be about 0.85 to 0.90.

The wavelength solution? It is necessary, as discussed earlier, to find a clear path; we had initially planned on using a pyrometer at 2.2 micrometers. From our standpoint, this is also an acceptable wavelength for our processes. Flame interference or soot in our furnaces is unimportant. The instrument manufacturer however had a unit at 0.95 micrometers.

D.P. DeWitt: Please comment on whether extensive use is made in industry of in-place targets like you have done.

P.D. Hunsucker: I do not know of any others. As a matter of fact, I have just received a patent on this target. In doing so, the patent examiner pushed many things and said that what I had was not new. In an extensive literature search, I found no similar devices except in steel mills; billet temperatures were measured at a reference place that was permanently positioned.

D.P. DeWitt: In the paper by Madding et al., as presented earlier here, reference is made to a two-target probe developed by Gulf R and D. Also Wayne Friesell of Infrared Scanning Services uses a target probe with the AGA Petroscanner. In response to a question from the floor, another firm indicated that they use a similar probe.

H.C. Hottel: I considered targets in the 1930's. In graduate student problems, I asked how to interpret furnace tube temperature measurements by having an auxiliary block between the tubes. This block had about the same view of the refractory and the flame as the tubes. The output had to be corrected to determine the temperature of the tube.

L.F. Albright, Purdue University: Concerning the targets, apparently you are making the assumption that the temperature of the target is uniform across the whole target. Yet based on my experience, the temperature on the surface of the target varies considerably. Yet the reading of an optical pyrometer will report an average. Would you comment on that?

P.D. Hunsucker: I have been making targets for two or three years. There was considerable design required to make a uniform temperature over the viewing surface. The air used for cooling the target is released on the backside of the target in a tangential manner so it will swirl around and promote uniform heat exchange. We had to do this because some of the pyrometers have a very good optics and they can look at a 25 mm dia. spot at 3m (1" dia at 10'); some are not quite that good. A tube is about 100 mm (4") diameter.

I prepared several pages of criteria for designing this target before I built one. Failure to do certain things result in a target that does not work. Thermocouple design also requires consideration of numerous factors including attaching them from the backside of the tube, drilling holes, and attaching the thermocouple wire discretely to the metal surface of the tube. Otherwise either heat is added to the junction or is taken away.

J.W. Friesell: Relative to the target, we have an extremely high-resolution instrument. When we look at our 50 × 50 mm (2 x 2 in.) target, we are probably looking at several hundred temperature points. On the target, we see a temperature distribution across it, and one of the critical parts of our technique is knowing where and when to take the reference temperature.

D.P. DeWitt: What kind of advice do you give people concerning looking at tubes on a near-normal or off-normal directions?

P.D. Hunsucker One problem that is really unanswered here is that we have noticed and realize there is reflected radiation. We do not know from where it all comes since the surface of the tube is sandcast and is very rough and pitted. With a smooth tube, one could say more concerning the angle of incidence. With a rough tube, it is more of a guess. One might integrate over the surface to obtain an average or equivalent angle. This approach may however not be very good.

D.P. DeWitt: Is your advice is to be as close to the normal as possible?

P.D. Hunsucker: For our measurements, we could not do them at a normal angle. So we measured the temperature of the wall and I estimated by several different means the average wall temperature. I have used the wall temperature just opposite the surface of the tube. Using the upper angle, I would integrate several points. Finally I obtained the most consistent data by integrating a wide area of the wall, i.e. the temperature of the wide area of the wall.

W.R. Barron, Williamson Corp. - Panelist

Like the other panelists, I was asked to look at certain areas. When we talk about accuracy, it is advantageous to listen to people having furnaces tubes and to learn what they can do. Let's also ask the question what can an instrument do. I think many process engineers, and other plant people, have some concerns about the accuracy of a radiometer and matters affecting the reading. My comments will be related to some made earlier. From an instrument point of view, generally instruments are designed and listed as having a calibration accuracy of ±1% at full scale. Most instruments are calibrated within that range, probably almost within 0.5% in order to stay within the 1% limit. The elements of calibration accuracy are the blackbody references; as indicated earlier, a group is establishing standards regarding not only blackbody references but also fields-of-view and other basic information. Another factor that affects the calibration accuracy of the instrument is the linearization electronics and also some component variables that

will occur in a design of a radiometer. Concerning repeatability or reproducibility of the instrument, several factors are sometimes significant. Does the calibration change as the instrument is taken from an air-conditioned office to the field, etc? Is ambient stability or long-term aging of importance? The size of the instrument is also important as operators generally like small instruments. It makes me nervous when people talk about "small is nice" and similar things since there are other factors for a radiometer besides it geometry.

From an application standpoint, I am very concerned about the field-of-view. My recommendation formula for a furnace tube is the field-of-view should be 1/3 the diameter of the tube. You do not want to fall off one side or the other side. So, if you have a 6 in. diameter tube, there should be a 2 in. field-of-view You need to check your field of view at varying working distances. If the field-of-view is 2 in. at working distance of 10 feet, at 30 feet the instrument may be reading other surfaces than those of the tube. It is important that people become quite aware of the field-of-view and optics. A certain lens system is required to provide good distant optics at certain spectral regions. Field-of-view measurement data is probably one of the weaker links of radiometry today. Some manufacturers say their field-of-view is within 67% of the energy; that is, 67% of the energy is coming within the specified source. So users should be aware of field of view requirements and test by looking through an opening x inches in diameter at certain working distances.

Concerning spectral filtering, mention has been made of 0.65, 0.95 and 2.2 micrometer wavelengths. Others use spectral filtering in and around 3.82 μm. Other windows are also important, but you want to see the filter. Is it a wideband filter, or a narrow-band filter? Someone can give you a 3.8 filter but if it is so wide it does not mean anything but if it is narrow it does mean something. It is important to know the bandwidth at 3.8 μm.

The comment earlier about operator training is excellent. I remember visiting a plant that had bought in an instrument earlier from us. The instrument now appeared battered, and I said "What have you done to it?" The plant supervisor was very concerned since a very expensive instrument had been purchased and it did not hold up very well. Obviously the instrument had been mishandled. An important aspect of using radiometers is operator training on how to use the instrument. Unfortunately, engineers often do not like to get involved with such training. The use of video is also very important. Much is to be gained by telling an operator why the instrument is being used, how to use it, how to aim the instrument, and general rules-of-thumb.

Concerning temperature distribution, I think Selas has a publication that indicates portions of the furnace tubes facing the flame are often 30-40 °F hotter than the opposite side. So when inspecting tubes in a furnace, I measure the front of the tube to get the maximum temperature. Measurements on the other side of the tube (on the side facing the wall) often indicate 20-30 °F lower. The radiant energy on the other side of the tube is much less due to the angle of the surface facing the radiant wall.

Concerning data collection, it is important to collect data accurately and to interpret it well in order to get effective results. As mentioned earlier, the furnace probe concept is very valuable relative to obtaining accurate results. The target as discussed earlier is often a section of tube, a couple of thermocouples are inserted in the wall of target, and steam or air is used to cool the target down to the approximate temperature of a typical furnace tube. Hence the environment for testing an instrument is comparable to a process tube. With this method one can periodically test the radiation instruments. Exxon uses such a device and so do other companies. Probably 2-4 times a year would be a typical time to run some tests to see how things are working, and to determine the degree of confidence for your operation. Blackbodies and other things can also be used. A target of the same material of construction as used in the plant in my opinion is a more realistic way to determine accuracy of a measurement as compared to just a blackbody reference. One may also get good blackbody reference, but it may be sensitive to flames or the optics may not be suitable.

In checking accuracy and performance of instruments, people often say one instrument is not reading the same as another instrument. How come they are not reading the same? Let me explain with an example. With a blackbody radiation curve and assuming wall temperatures of 2200 °F, furnace tube temperatures of 1700 °F, and emissivity of 0.9, the instrument filters at longer wavelengths more radiation from the wall than at shorter wavelengths. So, for instance, if you have filtered at 3.8 μm, the temperature error is about 45 °F high. At 0.8 μm, the error is in the order of about 80-85 °F high.

How do we interface the organizational structures between instrument manufacturer and the user? There is of course some direct contact, and technical societies including AIChE, ASTM, SPIE, ISA, and others providing important services such as communications, which is critical. Until two years ago there were no standards for radiometry. I believe that both ASTM and SPIE are working to develop industry standards. It is a slow political process but it is happening and there is benefit.

Confidence is developing in certain consulting companies but to date they seem to have limited field experiences which hurts them. They sometimes tend to be expensive, but it is an area where they need to initiate interest themselves. Those building the furnaces sometimes seem to take the attitude let the user worry about temperature measurement.

The instrument manufacturers tend to be basically small businesses. They are in general technically competent; they are and have to be application oriented but they have limited engineering. Considering the user of the instruments, their main operations are engineering oriented; they can define problems but as far as I can tell, they have only made limited commitments to solving some of these problems. There should be more participation by all in technical societies and more cooperation between the user and the instrument manufacturer. As instrument manufacturers, we have become involved in some industries where we have worked closely with the user of the equipment and its been a very successful relationship. Loaning an instrument or researching a little here or there are possibilities. I think the weakness of the whole situation is possibly not in engineering but in management. Management is not yet totally aware of what is needed, what is available, and what is not available. I think that the management levels of major corporations can learn much and take examples from other people who are successful in taking care of problems like this.

From an instrument manufacturer point-of-view, the goal should be to reduce the calibration/application error by half. I think that can be obtained. The world is getting closer and more actively involved with microprocessor electronics and the front end of instruments. Some thermographic portable equipment is advantageous for certain applications; multi-wavelength sensors are being developed; fiber optics are becoming more dependable; and things like laser infrared sensors may find applications. The latter two represent a concept in which an instrument would measure the reflectivity of the tube or surface, and then adjust the radiometer to read it properly since it has a built in calibration.

E.M. Emery: I agree with Mr. Barron one-hundred percent! We can measure the temperature if the client has the patience and the commitment to do it. Often when a client discovers the amount of effort required to make an accurate infrared temperature measurement, he backs away from it. Or, you go in and you break your heart to make the measurement for him and he says, "Oh, well that is really interesting. Thanks a lot. We will see you".

D.P. DeWitt: Fortunately some management is committed. Mr. Grantom from the Chocolate Bayou Plant of DuPont talked yesterday about a program that extended over 2-3 years and involved both education and calibration. Management in that case made a real commitment but Mr. Grantom and his associates likely helped along the way to "educate" management to a considerable extent. Contrast that with someone else saying how much do two of these instruments cost? Maybe they buy instruments but provide no training or educational program to the operators. Obviously there is little commitment in the later case and a serious lack of committment is evident.

W.R. Barron: We have worked with the Alcoa Company. Two or three years ago they could

not measure the temperature of aluminum billets accurately. We cooperated with them on a joint program. We obtained certain samples and did certain tests. They participated in everything we did. We have added to their thinking plus that of others, and we built an instrument that can measures aluminum temperatures quite accurately. We have since gone to other industries who have not made this committment. Some "Mickey Mouse around" and want to borrow an instrument for six months just to find something wrong with it. Where the management in Alcoa decided to build some needed equipment, everything worked out very well.

D.G. Damin: We keep criticizing management, but the engineering people at the plant must also be committed, to dig into the problem, and to convince their management of the needs and potential benefits. We can not hide behind management. We engineers must take some responsibility ourselves.

Dr. H.C. Hottel - Panelist

This stimulating two days has made me much more aware of the increasing number of blind spots that accumulate with time. Most of you will not understand that comment for another generation or two. Mr. Grantom described this morning how conscientious and informed a group must be to use good but simple instruments to assess furnace performance; this was a very important contribution. It is something from which most of us can learn; the main reason that many poor readings are made in furnaces is that people are not enough concerned about the problem. The man in charge must have enthusiasm and see that his staff does too. I was much impressed with his comments.

Three papers presented at this symposium have emphasized the importance of choice of wavelength and the effect on absorption coefficients. By proper choice of wavelength, it is possible to ignore the intervening gas when you want to receive the signal from only the surface or it is possible to suppress the background and accentuate the gas radiation output when you want to look at the gas. In regards to temperature measurements of the surface of a tube, I believe that I was on the beam 40 years ago, in suggesting the reading of a secondary target in addition to the readings on the tubes. From these sessions, you may conclude that instrumentation has almost reached its ultimate to measure tube temperatures.

In this symposium, there has been a strong focus on tube temperatures and almost nothing about gas-temperature measurements. There are however ways to measure the gas temperature and for certain furnaces including steam boilers, the gas temperature, and particularly the temperature gradients in the gas, are of significance. To do a first-class modelling job on a big steam boiler with an enormous path length through the gases, temperature distribution in the gas must be known. There is a phenomenon that must first be described however to make my next point. Let's consider a fluid with an emissive power which is linear in distance from the observation point to infinity; however infinity can be relative short if the absorption coefficient is finite. For such a gas, the instrument averages the temperatures of the gases. It can be shown that the temperature reported is that at one mean free path of a photon from the observation point. Mean free path of a photon is the inverse of the adsorption coefficient that has been discussed on several occasions in this meeting. So, an instrument looking at gases of varying temperature gives an apparent temperature existing one mean free path away. Meteorologists use this approach to determine the temperature profile of the atmosphere. They even have established the amount and position of water vapor layers in the atmosphere and know that the CO_2 is uniformly distributed.

In a furnace, there is however not a linear variation in emissive powers, so there is a major calculation problem. The approach used is to conduct monochromatic measurements at several wavelengths to obtain several absorption coefficients. The wavelengths need to be chosen with care. One wavelength needs to be associated with the mixture of CO_2 and water vapor. The ratio of CO_2 and water vapor in the combustion gases in the furnace is calculated based on the ratio of air and fuel to the burners and on the composition of the fuel. Another wavelength is picked to look at only CO_2. At a short wavelength, one looks at only the soot. As indi-

cated in my presentation yesterday, the absorption coefficient of coke varies with wavelength. One obtains a collection of absorption coefficients and apparent temperatures from which the temperature distribution can be inferred as a function of depth. It is quite a trick but worth doing; I predict the instrument makers will some day use this approach. Some developments have been made in Australia. M. Stewart, former head of Chemical Engineering in New Castle and a former student of mine, and one of his associates, Terry Wall, have specialized on the gas temperature patterns in furnaces. Their sophisticated models based on the zone method of calculating furnace performance provide information of temperature distribution in the combustion gases. The designer of a steam boiler wants to know the flux in the unit and how he can change it with burner design, with excess air, etc. He is less interested in the surface temperature which is usually rather low.

J.W. Friesell: Recently we were trying to get some temperature distributions in a flare and to measure the temperature at the tip of the flame, but with no experience. I obtained a thermogram using an imaging infrared scanner. We predicted from our calculations and calibrations that the hottest part of the flame would be $2300°$F. We were told that was the theoretical flame temperature.

H.C. Hottel: If that was the theoretical flame temperature, there was much excess air.

J.W. Friesell: It was a one-shot test and it made an interesting picture. Maybe we were lucky with the temperature, but it always intrigued me so I kept the picture. I will show it to anybody that is interested.

Quantum Logic Representative: Dr. Hottel, what spectral resolution is needed to get the accuracy desired, and how stable does the thermal distribution in the fire box have to be before meaningful data can be obtained?

H.C. Hottel: I visualize it as a technique to be worked on but I personally have not worked on it.

H.G. Semerjian, National Bureau of Standards: I would like to expand on Prof. Hottel's comments. Most techniques discussed at this symposium are not new techniques but are techniques being perfected for specific applications. I would like to call attention to some applications in aerospace-related areas, gas turbines, etc.

Germs of new ideas may be found for the petrochemical industry. Several techniques are being developed especially for gas temperature measurements. For example, the frequency-scanning techniques employed in satellites can be employed to determine the temperature and the composition distribution in the atmosphere. They are also being used for temperature measurements in combustion environments. There is the application of FTIR (Fourier transform infrared techniques) that gives similar information because one has both frequency scanning and spatial scanning. Two-dimensional arrays have been employed in a variety of applications involving both passive and active measurements. For example, a laser beam is employed to determine scattering. One can visualize the flow, and determine temperature and composition in such environments. Much more sophisticated techniques involving for example, CARS (Coherent Anti-Stokes Raman Spectroscopy) has been used in a gasifier —a full-scale, high-pressure gasifier. Very useful information has been obtained using such techniques. These techniques are sophisticated, and they are not necessarily applicable for everyday plant use. They likely will be very useful for diagnostics, specially by plant designers in order to improve their design criteria and to understand what goes on in a process furnace.

There are both passive and active measurement techniques, active measurements include laser-induced processes; passive measurements can utilize more detail in terms of frequency content of the radiation. New approaches are available and can be developed for future use in this industry.

H.C. Hottel: I am glad to hear those comments. I think that before one got as far as CARS, which I considered talking about yesterday but decided was a little bit too advanced, one should cover the ancient Kurlbaum method of having a blackbody background behind the gas. The instrument employed (either a total radiation, an optical pyrometer, or any other monochromatic instrument) is then adjusted until the temperature reading is the same when looking straight at the background or when looking at the background through the gas that you want to know about. When the two temperatures are alike, the background is at the gas temperature. The trouble is that a peephole is required on the backside of the furnace where you wanted to look in and have a control temperature background back there. That is quite a nuisance. So, what is known as the modified Schmidt Kurlbaum or modified Schmidt method is used. (It was named after H. Schmidt and not E. Schmidt). The Schmidt Kurlbaum method uses a refractory background instead of a controlled background. It works with a monochromatic instrument. It provides however an erroneous answer with a total radiation instrument because emissivities and absorptivities are both assumed to be alike. Of course, the total absorptivity of the flame for radiation from various points in the furnace depends on the temperature of the various point. That ruins the use of the Kurlbaum principle unless monochromatic radiation is employed. A whole additional symposium on gas temperature measurement is needed to cover the subject of gas temperature measurements.

D.P. DeWitt: At the ASTM-NBS symposium April 1984 (see ASTM STP 895) on applications of radiation thermometry, Dr. Semerjian discussed some very exciting technology. Although it is very complicated technology, it is moving from the laboratories into pilot plants. Who knows when it will be used in operating plants. We should probably be more sensitive to the newer approaches.

H.G. Semerjian: Within a week and a half, we will test a prototype unit in a recovery boiler in a paper mill. A highly corrosive solution which is the combination of lignin, caustic, and sodium sulfides is fired in order to recover the sodium sulfide (used in the digestion process). It is called a black liquor because it is like a tar, and it is viscous and sticky. Wavelengths must be selected very carefully. Emitters such as potassium are present and prevent blackbody emission.

Two wavelength measurements are made in order to deduce the temperature. Applications such as this can be utilized in industrial systems. Other applications, such as CARS, require more complicated instrumentation and more education. One really needs to be a Ph.D. physicist to understand the operation of that particular set up. Other techniques in between those are being employed in gas turbine applications. I think the petrochemical industry has much to learn from the gas turbine industry because they have been in the forefront of combustion science. Numerous techniques from the latter industry could be applied quite directly to the petrochemical industry and to combustion processes.

D.P. DeWitt: Dr. T.R. Todd of Exxon Research and Engineering Co. has agreed to speak about an interesting development at his company.

T.R. Todd, Exxon: An important parameter that has been omitted in this conference has been emissivity. Speakers during the last two days have estimated their emissivity from tables or possibly by going to the laboratory and making measurement. At Exxon, a laser pyrometer which is a combination laser reflectometer and a conventional infrared radiometer has been developed. Using the laser, a beam is aimed at the target. The energy reflected from the target is directed to the pyrometer. A measurement is made of the reflectivity of the target from which the emissivity of the target is calculated. This technique is of course limited to diffuse reflectors or Lambertian radiators; the furnace tubes being sandcast are quite good diffuse reflectors. We have developed a prototype instrument that has been used with good success in our Exxon plant for over a year. The reflectivity measurement has been coupled with the radiance measurement and the wall temperature measurement to make temperature corrections and to calculate an absolute temperature on the target tubes. We have not as yet licensed this technology, but we intend to.

D.P. DeWitt: Do you have any questions for Mr. Todd on this subject?

J.M. Soh, Arco Chemical: What is the incentive to make the emissivity measurement or the reflectivity measurement and how does one convince management that it is important?

T.R. Todd: By demonstration. We measured the emissivity or reflectivities of tubes samples in the laboratory using a laser. For a given sample we demonstrated that the emissivity was location dependent, i.e. it depended on where one targeted the tube. We have also shown in the laboratory that emissivity is temperature dependent. Based on some field data, it is also likely dependent on the composition of the tubes. As the tubes age, their emissivities change. I think it is important to make emissivity measurements of the tubes as they age at operating temperatures. Considering target probes as already described, they are only in the furnaces for a few minutes, or maybe an hour or two. They are not in the furnace for thousands of hours like the tubes, so they do not have the same aging factor as the tubes.

H.C. Hottel: Exxon sent me some time ago sections of tubes from a thermal cracking furnace. I made measurements of the emissivity by peening platinum couples into a razor blade crack on the surface of the samples. The emissivities varied greatly with the location on the section. Of course, the surfaces depended on what had happened in the furnace; the deposits on the surface were strange and wonderful.

J.M. Soh: Does the emissivity depend upon the wavelength and how accurate is it?

T.R. Todd: The emissivity depends on the wavelength and can be determined to probably 1-1.5%, in other words to within 0.01 emissivity units with the pyrometer. The unit is distance dependent because the scattering falls off as $1/r^2$. There is of course a maximum distance at which one can operate; it depends on the laser power. In terms of temperature accuracy and using all the equations and performing uncertainty calculations, the most uncertain part of the determination is estimating the wall temperature. That will ultimately limit accuracy more than any other component in the analysis. When determining wall temperature, one uses the correct reflectivity in the equation and the correct emissivity as measured in situ at furnace conditions.

D.P. DeWitt: In closing, I would like to thank the panelists and the audience for their spirited participation. The records of this discussion should be a valuable contribution to the field.

SYMPOSIUM SERIES

ADSORPTION

AEROSPACE

BIOENGINEERING

CRYSTALLIZATION

DRAG REDUCTION

ENERGY
Conversion and Transfer

Nuclear Engineering

ENVIRONMENT

FLUIDIZATION